Topical Conference: Decarbonization of the Chemical Industry through Electrification

Held at the 2024 AIChE Annual Meeting

San Diego, California, USA
27-31 October 2024

ISBN: 979-8-3313-1674-7

Printed from e-media with permission by:

Curran Associates, Inc.
57 Morehouse Lane
Red Hook, NY 12571

Some format issues inherent in the e-media version may also appear in this print version.

AIChE
120 Wall Street, FL 23
New York, NY 10005-4020

Phone: (800) 242-4363
Fax: (203) 775-5177

www.aiche.org

Additional copies of this publication are available from:

Curran Associates, Inc.
57 Morehouse Lane
Red Hook, NY 12571 USA
Phone: 845-758-0400
Fax: 845-758-2633
Email: curran@proceedings.com
Web: www.proceedings.com

TABLE OF CONTENTS

DECARBONIZATION BY ELECTRIFICATION: FROM ACADEMIC TO INDUSTRIAL PERSPECTIVE

DECARBONIZATION BY ELECTRIFICATION: THERMAL REACTORS AND REACTIONS I

DECARBONIZATION BY ELECTRIFICATION: SEPARATIONS

DECARBONIZATION BY ELECTRIFICATION: SYSTEMS AND POLICY

POSTER SESSION: DECARBONIZATION BY ELECTRIFICATION

DECARBONIZATION BY ELECTRIFICATION: NON-THERMAL REACTORS AND REACTIONS

DECARBONIZATION BY ELECTRIFICATION: THERMAL REACTORS AND REACTIONS II

293a Electrochemical Hydrogenation and Hydrogenolysis of Furanics Parameter Influences on Kinetics, Side Products and Carbon Footprint

Elizabeth J. Biddinger

There are significant opportunities to drive chemical reactions with electricity rather than heat as the increase in renewable electricity both reduces costs and decarbonizes the grid. Electrochemical reactions are amenable for this transition to electrically-driven systems and also offer opportunities to re-evaluate feedstocks and production scenarios. For example, electrochemical conversion of biomass-derived species at distributed biorefineries becomes more viable. There is still much that needs to be learned about the electrochemical reactions of biomass-derived species. For example, in the electrochemical hydrogenation and hydrogenolysis (ECH) of furanics the reaction parameters have a significant influence on the kinetics, selectivity and side product formation. The reactions are highly pH sensitive, with very low pHs required for hydrogenolysis to occur. However, at very low pH, undesired homogeneous side reactions also occur that consume the products. By pairing reaction mechanism and kinetics that we developed with homogeneous side reaction kinetics, models can be developed to better understand how the reactions compete in the reactors. These reaction rates can also be incorporated into larger carbon footprint models to better understand the influence of the reaction parameters. While highly dilute reaction streams minimize side reactions, the carbon footprint becomes very significant because of the separation needs downstream. Balancing reaction performance with separation needs becomes very important.

293b E-Heated Ethanol Dehydration for Sustainable Ethylene Production

Emre Turan, Robert Dieckmann, Michael Geske, Raoul Naumann D'Alnoncourt, Michael Bender, Benjamin Frank, Johannes Bode, Frank Rosowski

1. Introduction

The substantial reliance on fossil fuels to meet global energy demands poses a significant obstacle to achieving climate neutrality targets [1]. A profound shift is necessary, encompassing the adoption of renewable energy sources, advancements in heating technologies, and the utilization of regenerative feedstocks. Electrified catalytic reactors, leveraging technologies such as resistive, inductive, and microwave heating, offer a promising way to expedite this transition. While extensive research has delved into the resistive heating of electrically conductive wall reactors and structured catalysts in high-temperature applications, particularly in Steam-Methane-Reforming [2], reverse-Water-Gas Shift [3] and ammonia decomposition reactions [4], limited attention has been given to resistively heated catalytic fixed-bed reactors [5,6]. Implementing resistive heating in a catalyst bed consisting of granules/particles is technically challenging due to the need for physical continuity and control over the contact area between conductive particles to prevent hot spot formation [7]. Nonetheless, on industrial scale, catalytic fixed-bed reactors involving granules/particles remain prevalent [8]. Our goal is to electrify endothermal (bio)ethanol-to-ethylene reaction, optimizing control through an electrically heated fixed-bed approach for improved economic and ecological impact. The integration of regenerative feedstocks with energy-efficient heating technology aims to achieve sustainable olefin production.

2. Methods

A tube reactor was utilized to evaluate the bed resistivity of a physical mixture across the axial length of the bed. Variations in volume fractions and particle size were conducted to identify the optimal parameter for achieving a sufficiently conductive and reproducible bed. The catalyst bed comprised a homogeneous physical mixture of electrically conductive graphite particles and commercial alumina catalyst. Catalytic investigations, incorporating resistive heating, were conducted in a nearly adiabatic double-tube reactor with radial heat insulation, suitable for a 5 ml fixed-bed configuration. Further investigations involved scaling up the fixed-bed reactor from 5 ml to 500 ml capacity.

3. Results and discussion

Resistivity measurements showed that achieving the targeted bed resistivity requires a volume fraction of approx. 30% of conductive particles. The successful demonstration of resistive heating in an electrically conductive catalyst bed for endothermal ethanol dehydration showed that nearly full conversion of ethanol to ethylene could be achieved at 325°C. Comparing to the reference isothermal kinetics, the homogeneous temperature distribution is evident, indicating the lack of detectable hotspots. In contrast to the industrial adiabatic process [9], which uses 75 wt.-% water in the feed as a heat carrier to prevent significant axial temperature gradients, the feed in e-heated approach consists of pure ethanol, as illustrated in Figure 1. Consequently, significant heat losses in the heat carrier gas are inevitable. Additionally, resistive heating of the catalyst bed allows operation at a significantly higher weight hourly space velocity ($WHSV_{Ethanol}$) at a comparable gas retention time compared to the industrial process with a large amount of water, responsible for high gas velocity. The successful demonstration of scaling up from 5 to 500 ml was achieved, with axial temperature profiles being recorded for different feed compositions, covering both high and low water content. The nearly absence of water have shown lower temperatures in the catalyst bed, since kinetical inhibiting of water [10] and higher gas velocities can be eliminated with resistive heating of catalytic fixed-bed. Since water as heat carrier is not needed anymore, less energy input was observed.

4. Conclusions

In the conventional adiabatic process, the performance is influenced by the heat supplied through gas pre-heating, whereas in case of resistive heating, performance is kinetically controlled. Therefore, the reaction is driven to the catalyst's capacity limit, enhancing catalyst utilization. While the pre-heated process is not able to achieve the target performance in the absence of water, e-heated approach demonstrates the ability to operate in omission of heat carrier. Eventually, resistive heating can enable to run endothermal reactions emission-free, simultaneously increase productivity and process energy-efficiency by preventing heat losses in heat carrier. All in all, resistive heating can lead to intensify the process and decarbonize the heating of endothermal reactions.

References

[1] IEA (2021), Key World Energy Statistics 2021, IEA, Paris https://www.iea.org/reports/key-world-energy-statistics-2021, License: CC BY 4.0.

[2] S.T. Wismann, J.S. Engbæk, S.B. Vendelbo, F.B. Bendixen, W.L. Eriksen, K. Aasberg-Petersen, C. Frandsen, I. Chorkendorff, P.M. Mortensen, Science (New York, N.Y.) 364 (2019) 756–759.

[3] L. Zheng, M. Ambrosetti, A. Beretta, G. Groppi, E. Tronconi, Chemical Engineering Journal 466 (2023) 143154.

[4] A. Badakhsh, Y. Kwak, Y.-J. Lee, H. Jeong, Y. Kim, H. Sohn, S.W. Nam, C.W. Yoon, C.W. Park, Y.S. Jo, Chemical Engineering Journal 426 (2021) 130802.

[5] Soderquist et al., An improved process for the catalytic dehydrogenation of hydrocarbons, 1968.

[6] Y.R. Lu, P. Nikrityuk, Applied Energy 228 (2018) 593–607.

[7] L. Zheng, M. Ambrosetti, E. Tronconi, ACS Eng. Au (2023).

[8] R.J. Wijngaarden, A. Kronberg, K.R. Westerterp, Industrial catalysis: Optimizing catalysts and processes, Wiley-VCH, Weinheim, 1998.

[9] L. Roza, E. L. Faleiros, Method for producing olefines from ethanol fuel, 2014.

[10] J.F. DeWilde, H. Chiang, D.A. Hickman, C.R. Ho, A. Bhan, ACS Catal. 3 (2013) 798–807.

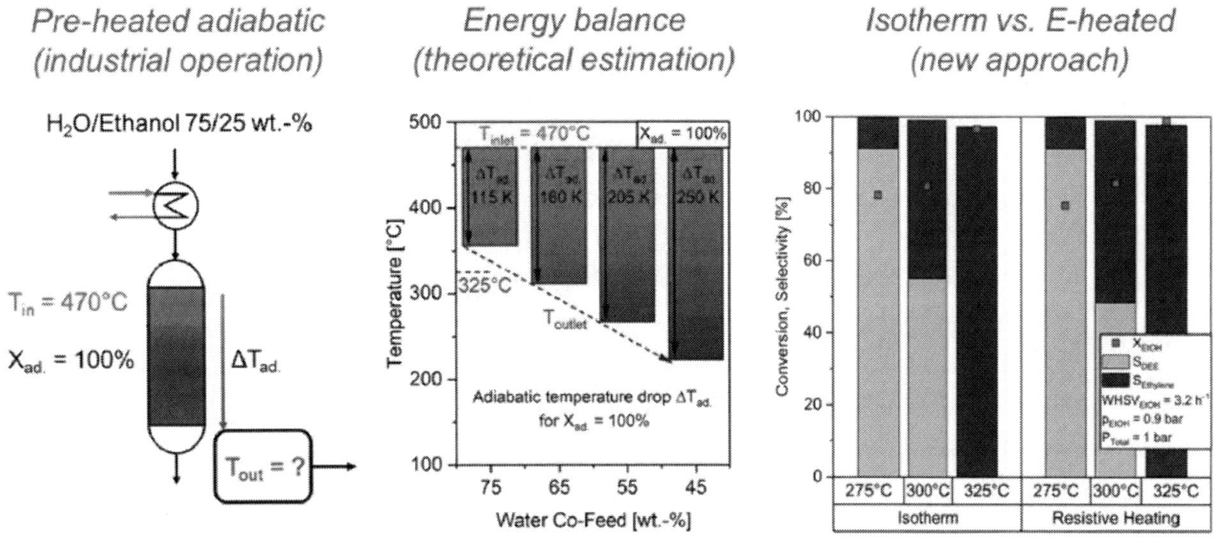

Figure 1: Comparison of the conventional adiabatic process and resistive heating on ethanol dehydration to ethylene

293c Towards a Circular Chemical Industry: Scaling up Plasma Technology from Lab- to Industrial Scale

Shahriar Mirpour, Yves Creyghton, Hans Linden, Taha Mana, Martijn Ruijzendaal, Maria Moreira de Azevedo, Tom Butterworth, Gerard van Rooij, Dirk van den Bekerom

Problem statement

Electrification of chemical industry

The chemical industry is responsible for a large fraction of global CO_2 equivalent emissions. Among the most polluting processes are naphtha cracking and hydrogen production for fertilizers. As significant contributors to global CO_2 emissions, they also present a large opportunity for reducing global carbon footprint. Naphtha cracking is an endo-thermic process that is heated partly by combusting the methane byproduct (15% by volume), resulting in a process that produces about 1 ton CO_2 per ton of olefins. Upon electrification of naphtha crackers, the process is nominally emission free when green electricity is used. Hydrogen for fertilizer production is produced through steam cracking of methane, followed by a water-gas-shift step. This process produces approximately 9 ton CO_2 per ton H_2 produced. Moreover, the hydrogen is first converted into ammonia, which is then oxidized to form nitric acid in the Ostwald process. In this process N_2O byproduct, which has a Greenhouse Warming Potential (GWP) of 273x that of CO_2, is formed and released into the atmosphere in significant quantities. Electrification of these processes would replace the fossil feedstocks as source of energy and thus eliminate most of the associated CO_2 emissions, which is urgently needed to reach the climate goals of 2030 and 2050.

Plasma technology

Plasma technology provides a unique pathway to electrify processes by applying electrical power directly to the process gas itself. The resulting plasma phase can be thought of as an electrical flame that can reach temperatures of thousands of degrees in a matter of seconds. Because the power is deposited locally in the process gas, the reactor vessel does not need to heat up, resulting in ON/OFF switching times in the order of seconds – ideally suited to follow the intermittent supply of renewable energy sources. Because heat is applied extrinsically it can be used to drive chemistry in any stable, low enthalpy molecule, allowing for truly circular processes. Moreover, the plasma reactor's electrodes can be made out of copper or steel and therefore do not inherently rely on any critical materials. The *Brightsite* consortium, a collaboration between knowledge institutes and industry (*Maastricht University*, *TNO*, *Sitech Services*, and the *Chemelot Campus*), works on scaling up plasma technology with the ultimate goal to scale up from lab scale to industrial scale. It is located on the *Chemelot* chemical site where it is surrounded by future end-users.

Methane plasma pyrolysis

Plasma pyrolysis of methane is the first plasma process being developed within the *Brightsite* consortium, which could have a significant benefit compared to more conventional methane valorization technologies. Upon electrification of steam crackers, a new purpose must be found for the methane by-product since burning it would nullify the emission reduction achieved in the electrification of the steam cracker. Plasma pyrolysis can be used to convert methane into acetylene or ethylene and hydrogen, opening up an avenue for valorization of the surplus methane. By producing polymer feedstocks as opposed to solid carbon, all of the co-products can be easily valorized, providing a significant economical advantage. This methane pyrolysis process is similar to the Huels process, which already employs plasma technology at industrial scale to convert methane to acetylene and hydrogen. The development goal of *Brightsite* is to improve the OPEX of the pyrolysis process by improving product yields and energy efficiency, as well as gaining a better fundamental understanding in the chemical pathways involved in the process. By focussing research on the direct production of ethylene, a significant improvement in energy efficiency can be attained by avoiding the exothermic hydrogenation of acetylene.

Further plasma processing opportunities

In addition to methane, plasma technology is developed for a range of other processes. Plasma processes can be competitive whenever stable feedstocks are involved or fast heating/cooling rates are required. One such application is the plasma fixation of nitrogen and oxygen to produce NO_x, a feedstock for fertilizer production. By producing NO_x directly, ammonia production can be omitted and the associated N_2O emissions eliminated, in addition to eliminating CO_2 emissions associated with the traditional steam reforming. By combining the pyrolysis of methane with nitrogen, a mixture of acetylene and HCN can be produced that, when catalytically converted to acrylonitrile, forms a pre-cursor for polyacrylonitrile. Additionally, conversion of CO_2 to CO and recycling of plastics is under consideration.

Methods

Fast quenching to control product selectivity

In the pyrolysis process methane can be converted to a range of hydrocarbon products. When considering product price and market size, acetylene and ethylene production is preferable over solid carbon, with ethylene the most desirable of the two. These hydrocarbons are formed through the Kassel mechanism, where hydrocarbons are subjected to successive dehydrogenation steps: $2CH_4 \rightarrow C_2H_6 \rightarrow C_2H_4 \rightarrow C_2H_2 \rightarrow 2C(s)$, balanced by production of molecular hydrogen. Due to the large spread in reaction rate constants of the different dehydrogenation steps, product selectivity can be tuned by quenching reactions to freeze kinetics once a certain product distribution is reached. Controlling residence times through appropriate application of the quench is thus a large focus in the research.

Scale-up strategy

Because a sequential scale-up takes considerable time and would not lead to maturity within the set timelines, a parallel scale-up strategy is pursued where different scales are developed simultaneously, each focusing on plasma technology of differing levels of maturity. Concretely, a 500kW pilot scale reactor will be constructed for investigation of the Huels process. Here the risk is minimized because the Huels process is already operating at industrial scale, demonstrating the process feasibility. A 50 kW "bench scale" reactor is being constructed for investigating opportunities to improve on the Huels process, still with a focus on ethylene production. The direct ethylene production, which is the significantly more challenging to achieve, is studied in lab scale reactors in the ~1kW range.

Experimental reactors

Most of our lab scale reactors are of the microwave type, where electricity is first converted in microwaves which are subsequently absorbed by the plasma to sustain it. The main advantage of this type of plasma is that it does not require electrodes and therefore has ample optical access. The setups are equipped with laser diagnostics such as Raman scattering for in-situ temperature and species measurements, and laser induced incandescence (LII) for investigations of soot growth. Downstream gas chromatography (GC) measurements measure the downstream composition. The 50 kW "bench scale" is a magnetically stabilized DC-arc, which better represents industrial scale DC-arc plasmas. It is equipped with a gas quench as well as a water quench to study the effect of different quenching strategies in a first scale-up step.

Results

Experimental

Downstream GC measurements recorded in the lab scale microwave reactors for a range of different parameters, provide insights into the ideal conditions for methane pyrolysis. In addition, it provides a means of tuning the conversion and selectivities by varying reactor conditions such as flow, pressure, and applied power. Raman scattering and LII were employed to gain a better understanding of the flow profiles in and around the plasma core, as well as the effect of radial temperature gradients.

Numerical modelling

In addition to the experiments, complementary numerical modelling was performed. 0D kinetic modelling was performed to understand the dominant chemical reaction mechanisms all the way to soot formation. These 0D models were complemented with computational fluid modelling (CFD) to estimate the typical flow and temperature profiles in the microwave reactor. Together the experiments, kinetic modelling, and CFD modelling paint a comprehensive picture of the pyrolysis process in the microwave reactor.

Conclusions

The chemical industry must rapidly be electrified in order to curb carbon emissions. Plasma technology is a promising candidate to electrify endothermic processes that require high temperatures, i.e. for driving chemistry in stable molecules to circularize product loops. Plasma technology is developed at *Brightsite* from lab to industrial scale, initially focusing on methane plasma pyrolysis. To scale up successfully, research is required over a large range of scales, with partners from knowledge institutes as well as well as industry. Following the successful scale up, these plasma processes pave the way for a green, circular chemical industry.

293d Electric Cracking Towers for Ethylene Production

Edwin Rodriguez, Rakesh Agrawal

Ethylene is an essential building block for countless modern products and services [1-3]. Global production surged to ~150 million tons in 2020 and is projected to grow to ~255 million tons by 2035 [4]. While Flame-heated Tubular Crackers (FTCs), illustrated in Figure 1, have dominated ethylene production for the last century [5-6], they face challenges with scalability, thermal management, and carbon emissions, accounting for ~0.6% of global releases [7-9].

In response, electrically heated crackers have re-emerged as a more sustainable option aligned with Net-Zero goals [10-11]. However, state-of-the-are designs, illustrated in Figure 2, encounter limitations akin to traditional FTCs, relying on heat transfer to the gas from the tube walls.

This study introduces Electric Cracking Towers (E-CRATOS), [US Patent App. 63/561,869, 2024] a novel reactor configuration, illustrated in Figure 3, designed to overcome the constraints of both traditional FTCs and current electric crackers [14-16]. E-CRATOS employs electric heating elements organized into stages along the reactor length, enabling superior thermal performance and scalability. As an electrically powered solution, E-CRATOS retains the benefits of decarbonization, reducing CO_2 emissions by up to ~95% compared to FTCs. The internal heating elements allow the use of larger diameter tubes for scaling, reducing construction metal and transportation weight. Additionally, strategic control of the heating elements facilitates temperature profile optimization, enhancing ethylene yield while reducing hot spots.

Our detailed Computational Fluid Dynamic (CFD) simulations, based on detail free radical kinetics and rigorous thermodynamics [17-19], show the potential for E-CRATOS to produce up to 4.2 times more ethylene than state-of-the-are electric tubular crackers for the same diameter, length, and surface temperature. Furthermore, by dynamically adjusting the heated length, E-CRATOS shows unparallel capabilities to withstand fluctuations in power delivery, making it an ideal technology for integration with renewable energy sources [14].

By fundamentally changing the reactor design, E-CRATOS represents an opportunity to achieve dramatic carbon emissions reductions while boosting production, bringing ethylene production one step closer to a Net-Zero Future [20-22].

References:

1. “Ethylene: The ‘World’s most important chemical,’” *American Fuel & Petrochemical Manufacturers (AFPM)*. Aug. 2022. [Online]. Available: https://www.afpm.org/newsroom/blog/ethylene-worlds-most-important-chemical

2. “Ethylene Oxide (EtO),” *EPA*. Environmental Protection Agency. [Online]. Available: https://www.epa.gov/ingredients-used-pesticide-products/ethylene-oxide-eto

3. K. Yamaguchi, H. Kishi, and T. Nagasawa, “On the use of Polyethylene Insulation in the Manufacture of Submarine Cables,” *IEEE Trans. Electr. Insul.*, no. 1, pp. 75–79, 1977.

4. “Global Ethylene Market Analysis,” *Research and Markets*. [Online]. Available: https://www.researchandmarkets.com/reports/5743468/global-ethylene-market-analysis-plant-capacity?utm_source=GNOM&utm_medium=PressRelease&utm_code=z75q4z&utm_campaign=1828184+-+Global+Ethylene+Market+Analysis+Report+2023%3A+Plant+Capacity%2C+Production%2C

5. H. Pierre and G. Curme, “PRODUCTION OF ETHYLENE,” U.S. Patent 1,460,545 (1923)

6. “Landmark for the birthplace of the petrochemical industry,” *American Chemical Society*. [Online]. Available: https://www.acs.org/content/acs/en/education/whatischemistry/landmarks/petrochemical-

industry-birthplace.html

7. H. Zimmermann and R. Walzl, "Ethylene," *Ullmann's Encycl. Ind. Chem.*, 2000.

8. S. M. Sadrameli, "Thermal/catalytic cracking of hydrocarbons for the production of olefins: A state-of-the-art review I: Thermal cracking review," *Fuel*, vol. 140. Elsevier Ltd, pp. 102–115, Jan. 15, 2015. doi: 10.1016/j.fuel.2014.09.034.

9. Z. Liu, Z. Deng, S. J. Davis, C. Giron, and P. Ciais, "Monitoring global carbon emissions in 2021," *Nature Reviews Earth and Environment*, vol. 3, no. 4. Springer Nature, pp. 217–219, Apr. 01, 2022. doi: 10.1038/s43017-022-00285-w.

10. Z. Chen, E. Rodriguez, and R. Agrawal, "Toward Carbon Neutrality for Natural Gas Liquids Valorization from Shale Gas," *Ind. Eng. Chem. Res.*, vol. 61, no. 12, pp. 4469–4474, 2022.

11. S. T. Wismann *et al.*, "Electrified methane reforming: A compact approach to greener industrial hydrogen production," *Science (80-.).*, vol. 364, no. 6442, pp. 756–759, 2019.

12. S. Sankaran and et al., "ELECTRICALLY POWERED FURNACES TO HEAT A FEED AND RELATED METHODS," WO 2023/006475 A1, Feb. 2023

13. S. Stevenson, "FURNACE INCLUDING HEATING ZONES WITH ELECTRICALLY POWERED HEATING ELEMENT AND RELATED METHODS." WO 2023/016968 A1, Feb. 2023.

14. R. Rodriguez-Gil, Edwin A.; Agrawal, "ELECTRICAL HEAT EXCHANGER AND REACTION TECHNOLOGY (E-HEART)," US Patent App. 63/561,869, 2024

15. R. Agrawal, Z. Chen, and P. Oladipupo, "ELECTRICALLY HEATED DEHYDROGENATION PROCESS," U.S. Patent 11,578,019 B2 (2023)

16. R. Agrawal and P. Oladipupo, "METHANE ETHANE CRACKERS," US 11,267,768 B2 (2022)

17. K. M. Sundaram and G. F. Froment, "Modeling of thermal cracking kinetics. 3. Radical mechanisms for the pyrolysis of simple paraffins, olefins, and their mixtures," *Ind. Eng. Chem. Fundam.*, vol. 17, no. 3, pp. 174–182, 1978.

18. A. Burcat and B. Ruscic, "Third millenium ideal gas and condensed phase thermochemical database for combustion (with update from active thermochemical tables).," 2005.

19. O. Olsvik and F. Billaud, "Modelling of the decomposition of methane at 1273 K in a plug flow reactor at low conversion," *J. Anal. Appl. Pyrolysis*, vol. 25, pp. 395–405, 1993.

20. J. KERRY and G. MCCARTHY, *THE LONG-TERM STRATEGY OF THE UNITED STATES Pathways to Net-Zero Greenhouse Gas Emissions by 2050*. 2021. [Online]. Available: https://www.whitehouse.gov/wp-content/uploads/2021/10/US-Long-Term-Strategy.pdf

21. J. Montgomery and M. Van Clieaf, *Net Zero Business Models: Winning in the Global Net Zero Economy*. John Wiley & Sons, 2023.

22. "Climate change: what the EU is doing." [Online]. Available: https://www.consilium.europa.eu/en/policies/climate-change/#:~:text=At least 55%25 fewer emissions by 2030,-Exactly a year&text=As an intermediate step towards,greenhouse gas emissions by 2030.&text=This goal is a major,of cutting emissions by 40%25.

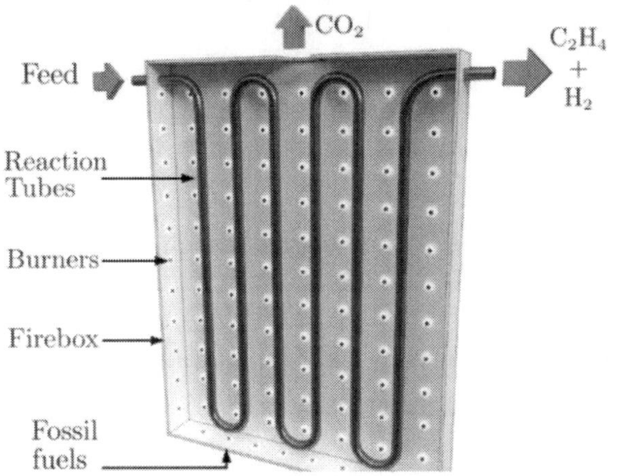

Figure 1. Illustration of a Flame-heated Tubular Cracker (FTC). External burners heat the reaction tubes to facilitate cracking.

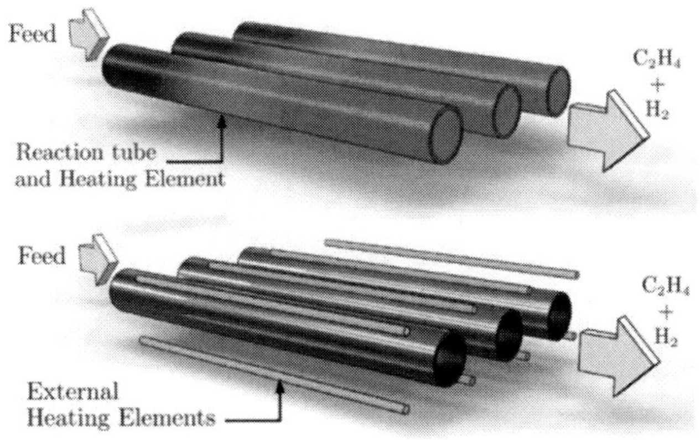

Figure 2. Illustration of state-of-the-art electrical tubular crackers.

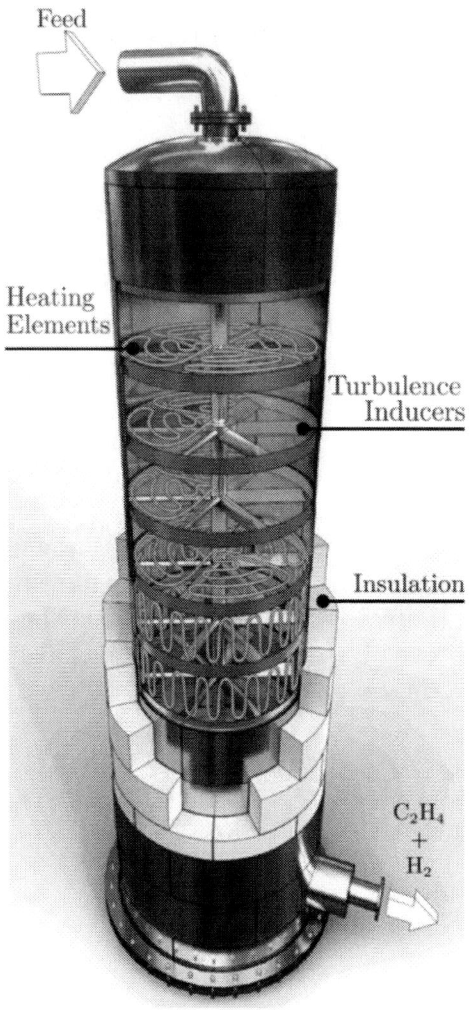

Figure 3. Illustration of an Electric Cracking Tower (E-CRATOS).

293e The Activation of Alkanes at Electrochemical Interfaces Using Real-Time Control of Potentials: Novel Avenues for Energy Storage and Sustainable Chemical Manufacturing

Marcel Schreier

Producing fuels and chemicals using renewable electricity holds the promise to enable a truly sustainable circular economy based on sustainably produced carriers of electrical energy and sustainably produced chemicals. The science which allows us to link electricity to chemical transformations, electrocatalysis, remains chiefly focused on the electricity-driven transformation of small inorganic molecules such as CO_2, H_2O, N_2, as well as the oxidation and reduction of alcohols. However, comprehensive electrification of society will require electrocatalytic reactions that can promote the central processes that sustain today's society: alkane transformations. In this presentation, I will show how fundamental understanding of the interfacial processes occurring in electrocatalytic reactions, combined with monolayer-sensitive differential electrochemical mass spectrometry, allows us to independently control the elementary steps (adsorption, transformation, desorption) of complex alkane transformation reactions, allowing for the room-temperature fragmentation of ethane into methane, as well as for enabling the extraction of the chemical energy stored in alkanes in the form of electricity. I will show how we can apply these technologies to expand the reaction scope of electrocatalysis, allowing for novel avenues in energy storage, chemical synthesis and plastics recycling.

563a Electrified Chemical Synthesis Enabled By Three-Dimensional Carbon Joule Heaters from Recycled Polyolefin Blends

Anthony Griffin, Jiachun Wu, Paul Smith, Gbadeoluwa Adedigba, Rui Shi, Yizhi Xiang, Zhe Qiang

With carbon neutral initiatives and the continuous decline of renewable energy source costs, electrification of heat for chemical reactions is a promising strategy to industrial decarbonization, improving sustainability for the future of our society. To address this urgent need, manufacturing of Joule heaters is an important research area; key design criteria include low-cost precursors, high customizability, and efficient heating performance. This talk will present an upcycling strategy to prepare three-dimensional carbon structures through additive manufacturing utilizing mixed polyolefin waste streams. Through a facile crosslinking and pyrolysis process, structured carbons can be manufactured on-demand with a vast geometric design space. Fabricated carbons exhibit excellent Joule heating performance, which can be employed as catalyst supports for the production of hydrogen through ammonia decomposition. Noteworthily, this electrified chemical synthesis exhibits improved reaction conversions compared to thermally heated analogs, while having the advantages of rapid heating and cooling. To demonstrate the generalizability of this process, we also extended this strategy to varied catalytic nanoparticles as well as prepared carbon from other recycled polyolefin blend compositions. This scalable process can address several existing challenges simultaneously, including plastic waste upcycling and industrial decarbonization.

563b Inductively Heated Swiss Roll Metamaterial Reactor

Connor Cremers, Chenghao Wan, Calvin Lin, Zhennan Ru, Dorothy Mantle, Ariana Hofelmann, Kesha Tamakuwala, Matthew Kanan, Juan Rivas, Jonathan Fan

With the need to globally scale a net zero carbon economy within the next few decades, it is essential to reduce greenhouse gas emissions from all sectors, including industrial chemical manufacturing. Electrified thermochemical reactors that are powered using intermittent renewable electricity sources, such as wind and solar, present a compelling technological route towards decarbonizing this industry while leveraging existing infrastructure. With electrification, power is delivered using electricity instead of the combustion of fossil fuels, eliminating the challenge, infrastructure, and cost involved in capturing and removing carbon emissions from combustion flue streams.

In a recent development, we proposed inductively heated metamaterial reactors, in which external electricity is converted to internal volumetric heat within a thermochemical reactor through the heating of a three-dimensional susceptor (i.e., the element that absorbs electromagnetic power). With proper multi-physics co-design of the susceptor, power electronics, and catalyst, it is possible to tailor and optimize physical properties including the coupling between the induction coil and susceptor, volumetric power delivery profile, effective thermal conductivity, and heat transfer from the susceptor to the catalyst. Notably, this can be exploited to reduce or even eliminate temperature gradients within reactors, improving catalyst utilization and system efficiency.

In a previous demonstration, we configured our metamaterial reactor to perform the reverse water gas shift (RWGS) reaction using a 38mm diameter conductive ceramic foam as a susceptor, and we showed that volumetric heating can eliminate radial thermal gradients allowing reactors to operate in ideal plug flow regimes. The reactor displays an experimental marginal efficiency of 86.6% upon increasing the GHSV from 1700 h^{-1} to 8800 h^{-1}, indicating a scaled reactor could achieve excellent power-to-reaction efficiency.

While these concepts show promise, susceptor design for the induction process in this regime fundamentally requires tradeoffs in desired properties. In this talk, we present the "Swiss Roll" resonant metamaterial reactor, which is a conceptually new framework for realizing structured inductively heated reactors. The concept is inspired by the electromagnetic Swiss roll, which was introduced to the electromagnetics community in 1999 and has been utilized in radio frequency metamaterial applications. The Swiss roll reactor consists of a sheet of metal that is rolled into a cylinder with a dielectric spacer or air gap between consecutive layers. The structure features a large self-capacitance and inductance and therefore supports a resonant magnetic mode at a frequency specified by the roll geometry.

The Swiss roll metamaterial reactor supports several distinct features. First, when inductively heated at resonance, the Swiss roll can be uniformly and volumetrically heated, distinct from non-resonant inductively heated reactors which feature some radial nonuniformity. Second, the Swiss roll reactor can be heated with almost arbitrarily high coupling efficiencies (i.e., relative heating of the susceptor versus that of the induction coil). By limiting dielectric and sheet resistance losses in the roll to achieve a sufficiently high quality factor, we have produced systems with 99%+ coupling efficiencies. Further, since this resonance is characteristic of the geometry and not the material, these efficiencies can be achieved at any design frequency. This elimination of coil losses is distinct from the non-resonant reactor case, where coil heating represented approximately half of the observed marginal losses in prior RWGS reactor studies.

Third, this intrinsically high coupling efficiency can be achieved at Swiss roll resonance independent of coil geometry or its spatial positioning. Relaxed constraints on induction coil geometry and position allow for thicker thermal insulation to be used in the reactor system, reducing thermal conduction losses. Multiple Swiss rolls can potentially be multiplexed with different resonant frequencies from a single coil to provide heating to different regions of the reactor in a simple and programmable manner.

Finally, the Swiss roll presents numerous viable manufacturing methods for large scale reactors. Depending on external design constraints, Swiss rolls can be additively manufactured, formed from sheet metals, rolled from flexible materials with a suitable high-temperature insulator, or subtractively manufactured via laser cutting or EDM. Catalysts can be washcoated onto the metal structure or loaded on supports between the layers as needed. Such scalability presents a major advantage over many non-resonant reactor concepts, which can't be manufactured large enough to be industrially relevant.

We demonstrate a lab scale Swiss roll reactor with a 96 mm diameter and use it to perform the RWGS reaction. We use DLMS 3D printed stainless steel to fabricate the Swiss roll, and powdered K_2CO_3/Al_2O_3 catalyst is readily added between the layers of the roll. In benchmark experiments with a similarly sized non-resonant metamaterial reactor, we show that the Swiss roll reactor operates with higher heating efficiencies due to its ability to accommodate thicker insulation and decreased coupling losses. Thermal profiling and multiphysics modeling display a high degree of radial temperature uniformity. The reactor shows near unity marginal efficiency upon process intensification, limited only by losses in the power amplifier.

563c Electrified Biogas Upgrading into Value-Added Chemicals

Yeonsu Kwak, Auden Jones, Weiqing Zheng, Dionisios Vlachos

High-temperature methane upgrading through electrification methods has emerged as a promising alternative to the traditional syngas process for converting methane into valuable products such as olefins, aromatics, and hydrogen. This study delves into the feasibility of various electrification techniques by comparing conventional, microwave-assisted, and continuous and pulse Joule-heated coupling of methane. The research emphasizes the significance of managing gas-phase reactions and carbon formation to ensure stable operation. Furthermore, the impact of introducing CO_2 and H_2O into the methane upgrading process is assessed, offering insights into the possibility of directly upgrading wet biogas. The outcomes of this study contribute to the advancement of cost-effective and environmentally friendly approaches for enhancing the value of biogas and natural gas, potentially reducing energy consumption and capital investments associated with producing fuels and chemicals from methane.

563d Steam-Methane Reforming on Joule Heated Ni-Coated Metal Wires

Elmer Ledesma, Meghana Sucharita Idamakanti, Praveen Bollini, Michael Harold

One potential low-carbon solution for SMR is electrifying the heating process using renewable power from sources such as wind and solar that do not emit GHG. Interesting approaches exist that utilize electricity to energize chemical reactors. Active research efforts are reported for e-SMR, although there are still no current industrial-scale processes. In this study we investigate the application of Joule heating of catalyst-coated, high-resistance FeCrAl wires as a potential method to electrify SMR.

The coating of Ni/ZrO_2 and Ni/Al_2O_3 onto 0.5-mm diameter FeCrAl wires involves slurry synthesis and coating steps. A wet impregnation procedure is used to synthesize the catalysts with two different Ni loadings (10 and 20 wt%). Slurries prepared with catalyst powders are uniformly deposited as a thin film onto coiled and straight FeCrAl wires. Joule heating of bench-scale reactor experiments are conducted by applying direct electric current through single wire and multiple wires. The feed gas is a mixture containing 5.4 vol% CH_4 with excess steam (H_2O:CH_4 = 3:1 – 4:1) molar ratio. This feed gas is kept at 190°C and atmospheric pressure and flows into a stainless-steel tube housing the wire. FTIR analysis is used to measure the effluent CO, CO_2, and CH_4 concentrations.

Major highlights: (a) SEM/EDS analysis reveal uniform coating of Ni/ZrO_2 or Ni/Al_2O_3 on FeCrAl wires; (b) CH_4 conversion is higher on Ni-coated wire compared to uncoated or ZrO_2/Al_2O_3-coated wire; (c) Discontinuous jump in CH_4 conversion at a critical applied power; (d) CH_4 conversion is higher on Joule heated wire compared to conventionally-heated wire; and (e) Stability experiments reveal that the catalyst-coated wires exhibited sustained activity, clearly demonstrating sustained SMR for several hours without catalyst deactivation. Several mechanisms are examined to elucidate the impact of Joule heating and the interesting differences between Joule and conventional heating through targeted experiments and reactor modeling.

563e Breaking Barriers: Reactive Study of a Revolutionary Turbo-Device for Defossilizing the Process Industry

Mike Bonheure, Rejish Lal Johnson, Elina Nauha, Kevin Van Geem

Introduction

Our world is facing huge challenges. Climate change, a problem created by humankind that goes far beyond national borders, requires a coordinated solution and demands a drastic change in our mindset and behavior. It is undebatable that our current society needs to develop an economic vision in which renewable energy and circularity thrive. In this respect, a crucial role is reserved for the (petro-)chemical industry, as it allows for the production of more sustainable products through changes in their processes and/or feedstocks. But before this chemical industry can become the key enabler of a sustainable society, it must become CO_2-neutral itself.

With a global energy consumption of 15 EJ of process energy annually, without considering feedstock-related energy, the chemical industry is responsible for a staggering 1.24 Gtonnes of CO_2, which is more than 18% of the total industrial greenhouse gas [1]. Even more surprising is that ~50% of this process energy consumption is related to only 7 base chemicals, which are ammonia, methanol, ethylene, propylene, benzene, toluene, and mixed xylenes. Therefore, by reinventing the production process of these 7 high-volume chemicals, there is potential to curb global greenhouse gas emissions by approximately 680 Mtonne on an annual basis, reducing global chemical and petrochemical process-related emissions by more than 54% annually [1].

The RotoDynamic Reactor™ (RDR™) developed by Coolbrook [2], offers the opportunity to drastically reshape the chemical skyline by reinventing how energy is transferred to the process fluid. Using meticulously placed turbomachinery blades, the fluid is accelerated beyond the speed of sound generating thereby shockwaves across which fluid properties like static temperature, static pressure, and density change almost instantaneously. Hence, gas heating becomes volumetric in nature rather than conductive/convective as in a conventional furnace.

To accelerate the development and widespread adoption of RDR™ within the (chemical) process industry, Computational Fluid Dynamics (CFD) simulations have been conducted. These simulations aim to provide a deeper understanding of the fluid dynamics within these groundbreaking turbo-devices. Moreover, by implementing a detailed pressure-dependent kinetic model into the CFD framework, in-depth insights are obtained about the product distribution within these devices. Pyrolytic coking rates in different zones of the reactor have for instance been analyzed, and the impact of chemistry on the reactor behavior has been studied through comparison with non-reactive results. Despite the massive potential of the RDR™ in the energy-intensive industry, this work focuses on the light-olefin industry, which is responsible for producing 5 of the 7 key building blocks - ethylene, propylene, benzene, toluene, and xylene – and therefore is the second-largest greenhouse gas emitter in the petrochemical industry.

Methodology

The complex geometry of the RDR™ is captured by utilizing a hybrid meshing approach containing both hexahedral and polyhedral cells. By leveraging the natural periodicity observed in the RDR™, only half the geometry must be modeled, which enables proper flux transfer across the domain through rotational periodic boundary conditions. In this work, a total of 18 million computational cells are used with adequate wall resolution ($y^+ < 5$).

The inherent transient nature of the RDR™ is captured through unsteady Reynolds Average Navier-Stokes simulations (URANS) with sliding mesh in the commercial software tool Ansys Fluent (v22R1). Given the highly turbulent flow within this device, an Explicit Algebraic Reynolds Stress Model (EARSM) proposed by Wallin and Johansson [3] is used for turbulent closure. This extended two-equation model accounts for a non-linear relation between the Reynolds stresses and the mean strain rate and vorticity tensors, thereby accurately accounting for anisotropy in the Reynolds stress tensor and enhanced secondary flow prediction due to improved pressure recovery in cases of strong adverse pressure gradients. Such gradients are notably observed within the RDR™ [4], thereby underpinning the selection for EARSM turbulence closure.

Although it is theoretically feasible to generate a reactive model by combining a microkinetic model with a suitable non-reactive model, practical limitations arise as the computational overhead increases more than linearly with the number of species in the gas phase. Hence, three chemistry acceleration methods have been utilized to allow a first-of-its-kind reactive simulation model of the RDR™, including a-prior model reduction, in-situ adaptive tabulation (ISAT), and chemistry agglomeration. Unlike pre-tabulation, which is prohibitive due to large table dimensions, ISAT stores thermophysical compositions of the gas phase only on the relevant composition space. When an already stored thermophysical composition is encountered, the output is retrieved from the table by passing the need for computationally demanding chemistry solution steps. Chemistry agglomeration further reduces computational load by binning cells with similar compositions, averaging their properties, and performing chemistry integration with the averaged composition, thereby reducing the number of costly chemistry calls.

Results and Discussion

The unique multi-regenerative design of the RDR™ utilizes a single electrically driven impeller section through which the fluid passes multiple times when flowing from the inlet toward the outlet. Each time the flow passes through the impeller, energy is added to the flow which is then dissipated into internal energy through a series of complex shock systems. Consequently, the RDR™ can increase the temperature above 1000 °C - see Figure 1(A) - in an order of magnitude faster than conventional fossil fuel-based furnaces.

Due to the intrinsically slow chemistry of steam cracking compared to the time scale of turbulent mixing ($Da << 1$), these reactions benefit from significant turbulent (micro)mixing. Hence, the diffuser within the RDR™ is specifically tailored to induce large-scale flow separation which progresses into the vaneless space where these large turbulent vortical structures will homogenize the flow-conditions which should favor the occurrence of steam cracking reactions. To study this, a pressure-dependent microkinetic model for ethane cracking is implemented, in which the added enthalpy will be used to enable and sustain the chemical reactions within the RDR™. This will result in overall lower temperatures within the RDR™ (see Figure 1(B)). Through this model, yields of economically interesting components can be assessed, see Figure 2(A), and the presence of pyrolytic coke-precursing components like benzene and butadiene can be observed, see Figure 2(B). This allows for an estimation of pyrolytic coking rates in these innovative devices [5]. Although there are inherent coking challenges in steam cracking processes, the RDR™ mitigates coke deposition through two key mechanisms: (1) direct mechanical energy transfer reducing wall temperatures, and (2) supersonic gas velocities promoting high shear stress, an order of magnitude higher compared to conventional steam cracking, thus inhibiting coke accumulation on the reactor surfaces. This translates to extended operational uptime and reduced maintenance concerns.

Furthermore, through a comparison of reactive and (hypothetical) non-reactive scenarios, it becomes evident that as the fluid undergoes chemical reactions, leading to changes in properties such as average molecular weight, its compressibility decreases. Consequently, this alters its tendency to achieve supersonic velocities. Moreover, the specific heat changes under the influence of the occurring reaction, which has a direct impact on the achievable temperature increase per stage. Indeed, the work applied to the fluid at a fixed rotational speed is directly proportional to the temperature increase per stage ($W = c_p \Delta T$). Even though chemistry significantly alters fluid properties throughout the RDR™, it displays remarkable resilience to these changing fluid conditions.

Conclusion

Addressing global challenges such as climate change demands a shift towards renewable energy and circularity, with the (petro-)chemical industry playing a pivotal role. However, before this industry can facilitate a sustainable society, it must become CO -neutral itself. The RDR™ by Coolbrook emerges as a transformative solution, as it reinvents energy transfer in chemical processes. Through Computational Fluid Dynamics simulations, this study achieves insights into the fluid dynamics and chemical behavior of the RDR™, particularly for light-olefin production. It is shown that this device enhances mixing, facilitating efficient steam cracking reactions while the resulting high shear stresses mitigate coke deposition. Its swift heating, above 1000 °C, which is an order of magnitude faster than current state-of-the-art, results in a highly selective production of economically interesting products like ethylene and propylene. Through comparative analysis between reactive and non-reactive simulations, it is shown that despite changes in fluid properties, the RDR™ displays remarkable resilience, underscoring its potential as a robust and sustainable technology.

References

[1] IEA, 'The Future of Petrochemicals Towards more sustainable plastics and fertilisers,' 2018. [Online]. Available: https://iea.blob.core.windows.net/assets/bee4ef3a-8876-4566-98cf-7a130c013805/The_Future_of_Petrochemicals.pdf

[2] Coolbrook. 'Electrification Solutions.' https://coolbrook.com/electrification-solutions/

[3] S. Wallin and A. V. Johansson, 'An explicit algebraic Reynolds stress model for incompressible and compressible turbulent flows,' *Journal of Fluid Mechanics,* vol. 403, pp. 89-132, 2000, doi: 10.1017/S0022112099007004.

[4] D. Rubini, N. Karefyllidis, L. Xu, B. Rosic, and H. Johannesdahl, 'A new robust regenerative turbo-reactor concept for clean hydrocarbon cracking,' *J. Glob. Power Propuls. Soc.,* vol. 6, pp. 135-150, 2022 2022, doi: 10.33737/jgpps/150550.

[5] K. M. Sundaram, P. S. Van Damme, and G. F. Froment, 'Coke deposition in the thermal cracking of ethane,' *AIChE Journal,* vol. 27, no. 6, pp. 946-951, 1981/11/01 1981, doi: https://doi.org/10.1002/aic.690270610.

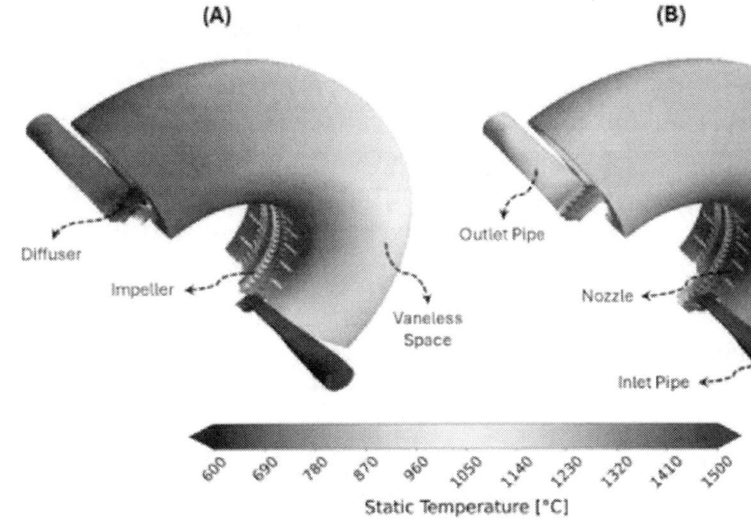

Figure 1: Instantaneous static temperature profile for ethane/steam mixture with a dilution ratio of 0.4 kg_{steam}/kg_{ethane}. (A): Non-reactive (hypothetical) simulation. (B): Reactive simulation.

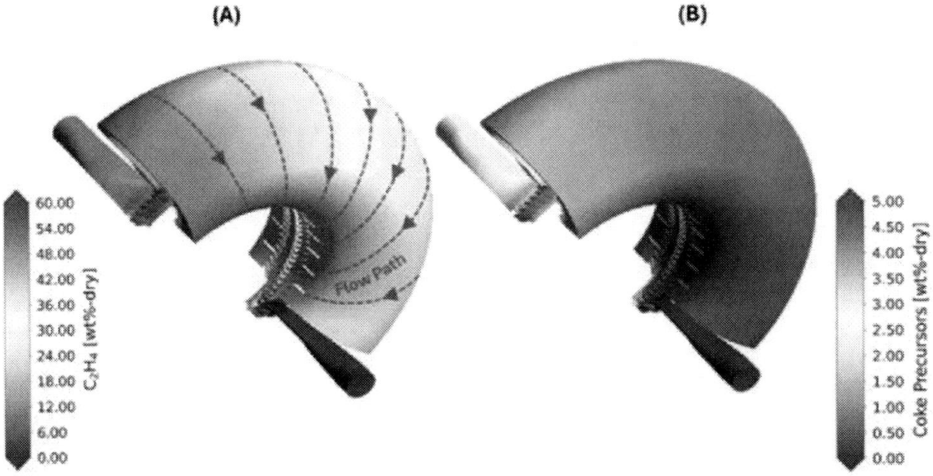

Figure 2: Instantaneous product distribution for ethane cracking with a dilution ratio of 0.4 kg_{steam}/kg_{ethane}. (A): Ethylene dry mass fraction. (B): Pyrolytic coking precursors, namely the combined dry mass fraction of 1,3-butadiene and benzene.

563f Stages and Kinetics of Poly(ethylene terephthalate) Depolymerization in a Ball Mill Reactor

Andrew W. Tricker, Yuchen Chang, Carsten Sievers

The deconstruction of post-consumer plastics into monomers provides the potential to both reduce the carbon intensity of chemicals and materials, as well as reduce the accumulation of environmentally harmful waste. Mechanochemistry, commonly performed in electrically driven ball mill reactors, offers a promising approach to chemically recycle plastics by avoiding the use of solvents and high reaction temperatures.[1] To this end, the progression and kinetics of the mechanochemical hydrolysis of PET under dry conditions at room temperature is investigated.[2] The rapid and complete depolymerization of PET is demonstrated with 20 min via ball milling with stoichiometric amounts of sodium hydroxide. Over the duration of the reaction, the yield of monomers increases steadily, while the molecular weight of the unreacted PET remains unchanged, indicating a reaction where PET completely depolymerizes within a local reaction environment, while chains outside this region are unaffected. A rapid increase in the depolymerization rate is observed when approximately 50% conversion of the PET, which corresponds to a transition of the PET powder into a homogenous wax. The transition to a wax increases accessibility of the PET chains, facilitating the more rapid hydrolysis. Finally, the rate of depolymerization is studied while adjusting milling parameter, specifically milling frequency, ball-to-powder ratio, and vessel temperature. From this, a unified kinetic expression to describe the conversion of PET is developed based on the milling energy dose.

References

1. K. J. Ardila-Fierro, J. G. Hernandez, *ChemSusChem* **2021**, *14*, 2145-2162
2. A. W. Tricker, A. A. Osibo, Y. Chang, J. X. Kang, A. Ganesan, E. Anglou, F. Boukouvala, S. Nair, C. W. Jones, *ACS Sustain. Chem. Eng.* **2022**, *10*, 11338-11347

563g Improving 1,3-Butadiene Selectivity with an Induction Heated Tandem System

Han Wang, Ben Ko, John Pham, Carlos L. Pueyo, Jianping Chen, Roel Sanchez-Carrera, Erdem Sasmaz

The electrification of process heating through induction heating (IH) has gained attention due to the possibility of direct usage of renewable electricity replacing heating through fossil fuel combustion. We have previously shown that IH can decrease the reaction temperature by 30 °C and reduce the energy consumption by 67% through the design of the catalyst bed configuration, demonstrating its ability to improve production and energy efficiency. In this work, we applied IH to the ethanol-to-1,3-butadiene (ETB) reaction using Co as the susceptor to provide heat inside the reactor. 1,3-butadiene can be produced in two reaction steps through the Ostromislensky process. Our results show that the first step of ethanol-to-acetaldehyde over Cu/Al_2O_3 catalyst can be lowered to 215 °C with IH, and the second step of ethanol and acetaldehyde-to-1,3-butadiene is preferred above 300 °C. Having each step perform at its preferred temperature in a tandem system can reduce energy consumption, produce fewer side products, and improve the selectivity to 1,3-butadiene. This can be achieved by applying IH as a targeted heating method; two temperature zones can be achieved in one reactor by controlling the amount of the susceptor.

Our results show that the tandem system can achieve 1,3-butadiene and butene isomers (C4 products) selectivities of 29.2% and 14.5%, while the selectivities are only 1.88% and 1.97%, respectively, over the ZrO_2/SiO_2 single catalyst bed with IH. The tandem system with IH improves the selectivity of C4 products by providing acetaldehyde as an intermediate to the second catalyst bed and controlling the temperatures separately for each reaction step. The selectivity of C4 products can be further improved by adjusting the reaction temperature, space velocity, bed configuration, and catalyst design. Our work has shown that the ethanol conversion of 62.9% and high C4 products' selectivity of 55.3% can be achieved at nine times lower power consumption. We show that IH can individually control reaction temperatures in ETB reaction for the first time, which can be a breakthrough technology for reducing greenhouse gas emissions and increasing product efficiency.

503a Selective Recovery of Fe and Rare Earth Element Concentrates from Tailings and Low-Grade Deposits

Duhan Zhang

As critical components for energy and defense technologies, the U.S. rare earth element ore reserves are estimated at ~13 million tons, but domestic production is currently limited to one operational mine residing on high-grade ore deposits in Mountain Pass, CA. While low-grade REE deposits are abundant, high-grade American REE deposits are sparse. Meanwhile, hazardous REE-containing tailings pile up at the economic and environmental cost, and it is estimated that REE value locked within e.g., red mud, is $796 per ton. However, such low-grade deposits or mining tailings are challenging to access because REEs are usually locked within a mineral matrix, typically iron oxide, that renders the ore intractable to mechanical or chemical separation. Here, we propose a process for recovering REEs from low-grade deposits and tailings by electro-metallurgically removing iron oxide, hydro-metallurgically removing aluminum salt, and separating insoluble rare earth oxides. Our strategy produces not only refinable REE concentrates but also emissions-free iron metal and supplementary cementitious materials as co-products for the steel and cement industries, respectively. Our technology simultaneously addresses the challenges facing the long-term demand of REEs as well as mitigates the challenges surrounding the management of industrial and mining waste. Compared to existing methods, our technology offers a zero-waste closed-loop process to maximize the extracted product value and minimize resultant waste. The developed process eventually realizes net positive returns, mitigates the growing environmental tolls of industrial waste, and improves the sustainability of domestic supply chains.

503b Improved Phosphorus Precipitation from Wastewater Using Pulsed Potential Waveform Approach

Ruhi Sultana, Lauren F. Greenlee, Michael Janik

Water recycling is a part of agricultural water resource management as a result of the growing world population and the anticipated intensification of climatic variability and fluctuations in dependable water supplies. Furthermore, valuable nutrients like phosphorus that could be recovered and reused to produce food (as agricultural fertilizer) are lost to wastewaters. Struvite has demonstrated potential as a substitute supply of phosphorus for fertilizer in agriculture. Studies conducted on row crops indicate that struvite (MgNH4PO4.6H2O) is a feasible fertilizer compound when compared to conventional fertilizers. Using an alkaline electrochemical reactor with a magnesium alloy anode and a stainless steel cathode for reducing water to hydrogen gas and hydroxide, our research aims to recover struvite from wastewater. The magnesium electrode that is used to produce magnesium cations in solution and precipitate struvite becomes seriously contaminated over time with subsequent reactor batches. This raises energy requirements and lowers fertilizer recovery for struvite. This research makes use of a dynamic, pulsed voltage waveform at the magnesium electrode surface producing a stress-strain-like effect that prevents or reduces the quantity of struvite precipitate that would adhere to the electrode. When compared to a constant voltage approach, an aqueous solution of ammonium dihydrogen phosphate (NH4H2PO4, 10 mM) at a neutral pH showed a 60% increase in divalent magnesium cations and a 40% increase in phosphate recovery as struvite.

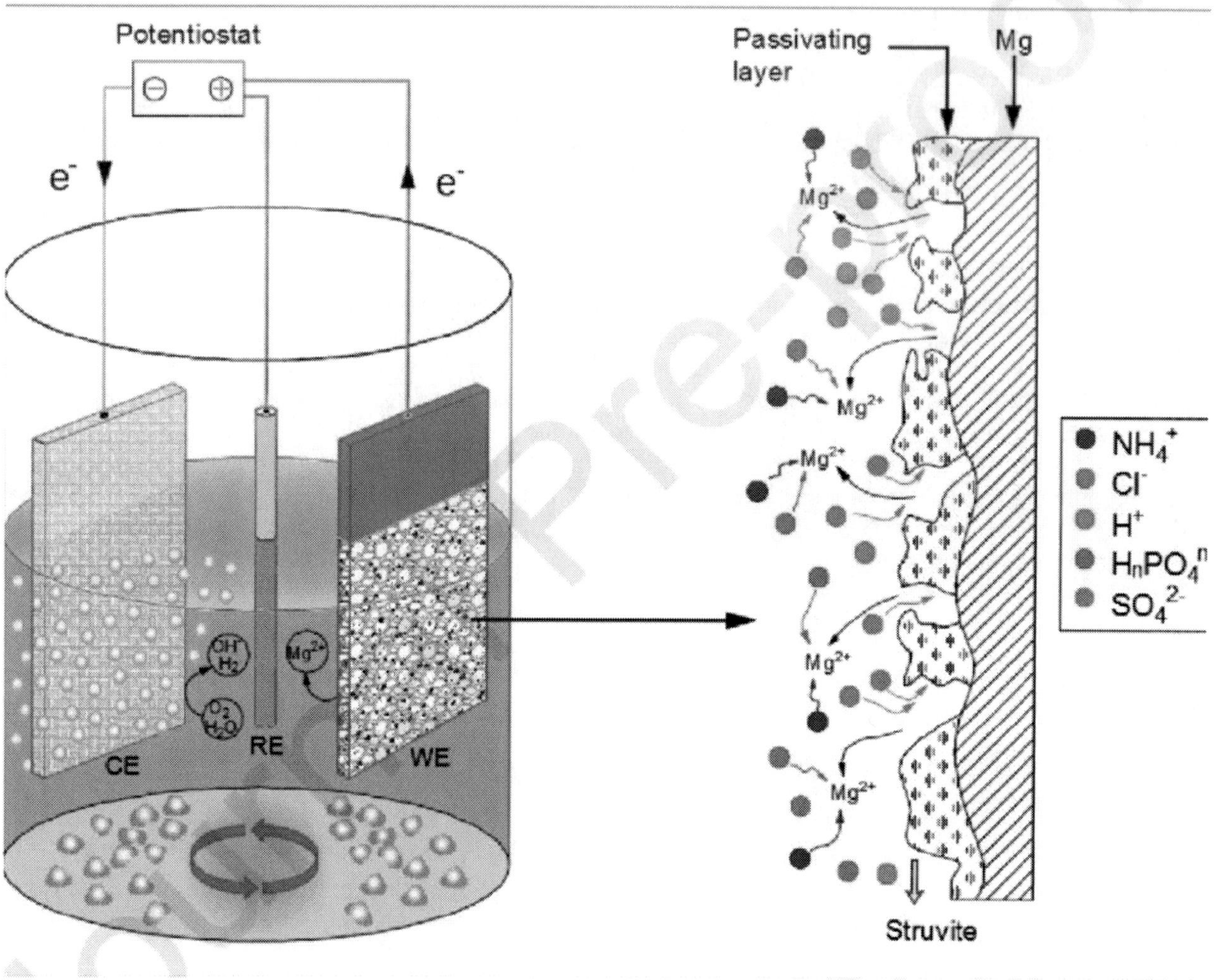
Potentiostat
e-
e-
CE
RE
WE
Passivating layer
Mg
Mg2+
NH4+
Cl-
H+
HnPO4n
SO42-
Struvite

503c Electrifying Membranes: Localized Joule Heating of Laser-Induced Graphene Membranes for Desalination of Hypersaline Brines

Mahdi Malmali, Hasib Mahbub, Mohammad Saed

Membrane distillation (MD) is an emerging desalination technique that can treat hypersaline brine to produce pure distillate. It is a thermally driven membrane separation process, in which the partial vapor pressure difference between the feed and the permeate sides leads to the water vapor molecules permeation through a hydrophobic microporous membrane. MD has received substantial attention for desalination of hypersaline brine (>75000 ppm), owing to the independence of mass on the feed salinity. However, MD's widespread application is yet hindered by low energy efficiency and scaling. In this presentation, I will report a novel, single-step fabrication method to develop electroconductive graphene membranes. The surface of these membrane can be heated using Joule heating. Thus, we can apply the heat at the interface of the mass transfer. This approach promises potential improved energy efficiency and reduced scaling by mitigating concentration and temperature polarization.

The electroconductive membranes developed in our work are fabricated by laser irradiation of a polyethersulfone (PES) substrate in raster mode lasing, which converts the surface of the PES substrate into a stacked multi layer of graphene sheets, also known as laser-induced graphene (LIG). Upon formation of the LIG on the membrane surface, it was coated with a thin SiO_2 @ polydimethylsiloxane (PDMS) layer to obtain hydrophobic surface for MD application. PES-LIG/SiO_2-PDMS membranes prepared were characterized to confirm the formation of graphene and protective hydrophobic coating. Results indicate the formation of a superhydrophobic (152°) surface on the PES-LIG membrane. For surface heating with the Joule heating technique, direct (DC) and alternating current (AC) were used with different input power ranging from 0.5 to 10W. The thermal response of PES-LIG membranes was recorded with an FLIR camera to find the optimum heating parameters (power and frequency for AC) to evaluate MD performance. Surface temperatures in the range of 50-350 °C were recorded for 0.5-10 W input power. The highest average distillate flux achieved was 8.74 $L.m^{-2}.h^{-1}$ (10W and 60 Hz) with AC and 6.34 $L.m^{-2}.h^{-1}$ (10W) with DC Joule heating. For both cases, more than 99% rejection of NaCl was observed, indicating the successful performance of the electroconductive PES-LIG membranes to sustain the mass transfer interface on the MD feed side. Our observation concluded that AC would be preferable over DC as polarity switching in AC significantly reduces electrode corrosion, which is essential for the long-term operation of MD.

503d Techno-Economic Analysis of Electrification for a Deep Decarbonization of the Industry

Frederick de Meyer

There are three main low carbon solutions for a deep decarbonization of the industry: carbon capture and sequestration (CCS), electrification with green power or using alternative low carbon fuels. We present a techno-economic comparison between the three options.

In this talk we will mainly focus on CCS. One of the main challenges of carbon capture is the high energy demand of the process. This is for example due to the use of heat for amine regeneration and reclaiming in an absorption-based process or cold in a cryogenic process. To avoid extra CO_2 emissions related to the production of heat or cold it might be interesting to electrify the carbon capture & conditioning process with green power. In this work we compare a classical and a fully electrified carbon capture & conditioning unit for different industrial applications (refinery, CCGT, LNG) for the three most mature industrial CO2 capture technologies: absorption with an aqueous amine, oxycombustion, and cryogenic separation (if applicable). We identify the key parameters determining the economics allowing for a quick decision. From an economic point of view, the main challenge of CCS is to build a strong business case, which is mainly based on the cost of avoided CO_2 emissions.

In a second step we compare CCS with full and partial (hybrid) electrification of the industrial plant with renewable or low carbon power. The business case is often much stronger because it includes not only avoided CO_2 emissions, but often also the sale of the hydrocarbon fuels that are being replaced. There are, however, serious technological challenges in the large-scale electrification of some equipment. Full and partial electrification and CCS are compared for typical onshore and offshore oil and gas plants, including LNG plants.

To conclude, CCS and electrification are compared to the use of an alternative, low carbon fuel. Examples are hydrogen and biogas. Similar to electrification, the business case is often much stronger due to the sale of saved hydrocarbon fuels. Alternative fuels are, however, limited in availability (biogas), or much more expensive (hydrogen) compared to usual fuels. Moreover, the technological maturity to produce and transport hydrogen at a large scale is still a challenge. However, hydrogen might become more advantageous compared to electrification if the process requires significant temperatures, for example, in a petrochemical plant, a furnace to crack hydrocarbons to produce monomers for the polymer industry.

The aim of the study is to identify the key parameters to guide us in the first selection of the optimal deep decarbonization approach. The comparison is performed both for greenfield and brownfield applications.

503e Engineering Processes and Materials for Downstream Purification of Products of Electrochemical CO_2 Reduction for an Ethylene Producing e-Refinery

Akriti Sarswat, David Sholl, Ryan Lively

Rising CO_2 emissions are driving significant investments in low-carbon and zero-carbon fuel production strategies. Decreases in renewable electricity costs have focused attention on materials and technologies to electrochemically reduce CO_2 into small molecules such as C_2 hydrocarbons. While there has been significant research on the electrochemical cell design and reactors for the system, there have been few studies, however, on the separation systems that will be eventually required to supplement the system. The presented work aims at strategically developing downstream separation process flows which significantly improve the overall conversion of CO_2 in these reactions without changing the reactor. Further, we identified major bottlenecks in adsorption-based separation processes. The presented work includes discovery and development of metal organic frameworks for these separations to further improve energy requirements in the downstream refining.

503f Electrochemically Mediated CO2 Sorbent Regeneration: Moving Beyond the Cu-Diamine System

Jonathan Boualavong

Electrochemical carbon capture has emerged as a viable alternative to the temperature-swing CO2 capture process due to its lower energy demands and greater potential for electrification. This process uses reducing and oxidizing conditions as analogs to the low and high temperatures of used in temperature-swing capture for sorbent regeneration. Electrochemically mediated amine regeneration (EMAR) has been shown to be among the most promising because it can operate at higher concentrations. In the EMAR process uses metal ions, typically copper, to disrupt the N-C carbamate bond of the amine sorbent by competitive inhibition, forming a Cu-N bond and releasing CO2. The metal ion concentrations are controlled electrochemically by electroplating and corrosion reactions to shift between a CO2-capturing state with low metal concentrations and a sorbent-regenerating state with high metal concentrations.

To date, the EMAR process has only been studied using diamine sorbents due to the strong metal-amine coordination achieved by the chelate effect and the established amine-CO2 chemistry. Additionally, experiments to date have prioritized energy consumption as its sole optimization metric, with limited concern for the CO2 absorption rate. In this study, we extend the model to electrochemically mediated sorbent regeneration (EMSR), account for non-amine sorbents that can capture CO2 in the (bi)carbonate forms via pH modulation. We have used a computational aquatic chemistry model of the system to simulate how the system's energy demand and CO2 absorption rate will change as the sorbent changes, studying the process across the breadth of theoretical bidentate ligands with amines and carboxylic acids as nucleophile sites. We selected this range of compounds to maintain the strong chelate effect that produces the Cu-CO2 competition for the sorbent, using carbonates to represent pH-modulating sorbents without carbamate formation capability. We find that the optimal pKa values for minimizing the energy demand of the system tend towards the pKa values of carbonic acid and bicarbonate, similar to our previous studies of the electrochemical pH-swing method. Amino acids can achieve energy demands and rates similar to that of the diamine system, opening up the opportunity for more water-soluble and environmentally benign amines in the EMSR process. While dicarbonate ligands are capable of CO2 capture in this system, the pKa values and metal coordination strengths achieved by dicarbonates lead to greater energy demands. Current work is ongoing to validate the results experimentally and explore other metal-ligand chemistries.

503g Scalable Nanoporous Networks for Carbon Capture Via Solid-State NMR Spectroscopy

Haiyan (Christina) Mao, Yi Cui, Jeffrey A. Reimer

We demonstrate new sustainable, solid-state, polyamine-appended, cyanuric acid-stabilized, melamine nanoporous networks (MNNs) via dynamic combinatorial chemistry (DCC) at the kilogram scale towards effective, scalable, recyclable, and high-capacity CO_2 capture. Our solid-state MNNs solves the long-standing problem of solvent degradation, equipment corrosion, and environmental leakage faced by traditional solvent methods with cheap solid raw materials (melamine and formaldehyde) through a facile synthesis method. Polyamine-appended MNNs reaction mechanisms with CO_2 were elucidated with double-level dynamic combinatorial chemistry. 2D heteronuclear chemical shift correlation (HETCOR) nuclear magnetic resonance (NMR) spectroscopy was performed to demonstrate the interatomic interactions between H and C atoms in the CO_2 reaction species at an atomic level. With quantitative NMR and density functional theory (DFT), we identified the reaction structure during CO_2 chemisorption, where ammonium carbamate pairs and a mix of ammonium carbamate and carbamic acid are easily distinguished. The coordination of polyamine and cyanuric acid modification endows the networks with high adsorption capacity (1.82 mmol/g at 1 bar), fast adsorption time (less than 1 min), low price ($40/ton), low regeneration energy (53 kJ/mol), and extraordinary stability to adsorption-desorption cycling by simulated flue gas. This work creates a general industrialization method towards sustainable CO_2 capture using networks (e.g., polymers, covalent organic frameworks, and metal organic frameworks) via atomic-level and dynamic combinatorial chemistry design strategies with precise mechanism studies.

357a Leveraging Hourly Variations in Electricity Pricing for Optimal Retrofit Electrification of an Oil Refinery

Sampriti Chattopadhyay, Rahul Gandhi, Ignacio Grossmann, Ana I. Torres

The chemical industry relies heavily on fossil fuels, presenting a significant challenge in meeting decarbonization goals. Despite the availability of technological solutions, the industry often hesitates to transition due to the value tied to existing infrastructure. To address this, we developed a comprehensive decision-making framework tailored for oil refineries. Our framework prioritizes the decarbonization of process heating and hydrogen production, leveraging advanced computational methods to identify the most cost-effective retrofitting plans.

Our preliminary research revealed that in the short to medium term, carbon capture technologies are preferred over electrification-based alternatives, unless there is a substantial decrease in electricity costs. Additionally, while the introduction of CO_2 taxes does incentivize the adoption of carbon capture technologies, it does not necessarily accelerate the adoption of electrification.

In the current work, in order to further enhance the viability of electrification solutions, and motivated by the general area of demand side management [1], we have investigated the effect of variability in electricity prices on an hourly basis, alongside exploring different storage options. Our findings indicate the potential for earlier adoption of electrified boilers and electrolytic hydrogen production when leveraging variable electricity prices.

In our results, we found hybrid technologies to be promising, such as combining electrified boilers with natural gas boilers, to optimize the cost of operation amidst hourly varying pricing scenarios. Additionally, leveraging storage options for hydrogen allows for increased flexibility, enabling higher production during periods of lower electricity prices.

To address the complexity of hourly electricity price variations and storage decisions, we adopted a representative days/weeks approach. By carefully selecting the number of representative time frames, we ensure that our model captures the full range of potential scenarios while maintaining computational tractability.

Moreover, tailored Benders decomposition is used to solve the optimization problem. By decomposing the problem into smaller, yearly subproblems, we enhance computational efficiency by parallelization without sacrificing accuracy. We add cuts to ensure that the subproblems are consistent with each other until convergence is achieved.

Finally, we briefly discuss how the proposed MILP models could be extended with stochastic programming techniques to explicitly account for uncertainties in the parameters of the MILP models.

Reference

Zhang, Q. and I.E. Grossmann, "Enterprise-wide Optimization for industrial demand side management: Fundamentals, advances and perspectives," *Chemical Engineering Research and Design* **116**, 114-131 (2016).

357c Designing a Holistic Systematic Framework to Ensure Effective Policies for Electrification in the Chemical Industry

Shayan Niknezhad, Efstratios Pistikopoulos

In the face of escalating climate change concerns and tightening global carbon emission regulations, the chemical industry stands at a critical stage where the urgency of managing carbon is not only desired but also mandated. Carbon is a valuable atom and the question is how to manage this molecule. The path to achieving this lies through the adoption of electrified chemical processes based on any raw feedstock powered by renewable energy—a transformative approach that cuts across traditional boundaries and merges technological innovation with policy reform. This holistic strategy aligns with the statement of our collective desire for a sustainable future and guides our actions towards responsible energy production, distribution, and consumption.

To quantify the decarbonization and electrification of the chemical industry, we must develop new methodologies and frameworks that consider the entire lifecycle of chemical processes. This holistic view will enable us to accurately measure the impact of our efforts and guide our actions in the most effective direction. Energy Attribute Certificates, including Renewable Energy Certificates, play a crucial role in documenting the generation and purchase of electricity from renewable sources. These certificates are vital in assessing the emissions associated with the production of clean products such as hydrogen, a leading candidate for the energy carrier in the decarbonized chemical industry.

This analytical process extends to integrating the complex interplay of politics and economics, systematically examining the efficacy of existing policies and identifying the most advantageous policy interventions. This entails a sophisticated analysis that combines direct costs with indirect costs (or hidden costs) and employs lifecycle assessments alongside hypothetical 'what-if' scenarios. By doing so, it is possible to craft well-informed policy strategies that not only advocate for environmental sustainability but also consider economic viability and political feasibility.

Electrification of the chemical industry encompasses a multifaceted approach. The focus lies on identifying and implementing electrified reactions and separations, fostering the design of new, more efficient processes, and ensuring seamless grid integration. Such innovation is necessitated by the policy landscape, which encompasses legislation, international treaties, and incentives for investment, leading to the creation of guidelines for energy conversion and supportive expenditures in the form of incentives, taxation, and credits.

To drive the chemical industry's shift towards decarbonization, policies and systems must prioritize the development and deployment of electrification in processes at varying temperature sable to support demonstrating and deploying pilot scales for electrification of selected low-temperature (below ~300 °C) reaction and separations units, high-temperature (above ~300) reactors for endothermic chemical conversions, and developing/demonstrating tools for identification of electrification opportunities for individual processes/plants, both existing and new.

In the U.S., the regulation of GHG emissions, including Scope 1 and 2, is underpinned by various federal laws and regulations. The most significant is the Clean Air Act (CAA), which, through mechanisms such as the New Source Performance Standards (NSPS), sets limits on emissions from new and modified sources in specific sectors. This legal framework establishes a robust baseline for addressing emissions directly associated with the chemical industry's operations.

However, the challenge of Scope 3 emissions—which encompass all other indirect emissions that occur in a company's value chain, such as those from the transportation of goods, business travel, and waste disposal—remains less directly regulated. These emissions are not governed by federal laws in the same stringent manner as Scope 1 and 2 emissions. Management and reporting of Scope 3 emissions are largely propelled by voluntary corporate sustainability practices, frameworks like the Greenhouse Gas Protocol, and the evolving demands of investors and consumers for greater corporate responsibility in mitigating climate change impacts. State-level initiatives also significantly contribute to the regulatory landscape, with several states adopting GHG emissions reporting requirements, cap-and-trade programs, or emission reduction targets surpassing federal mandates.

Who is to implement these changes? The responsibility lies with a coalition of stakeholders, including industry leaders, policymakers, technologists, researchers, laboratories, think tanks, and the workforce, all geared toward transforming the industry. The collective actions aim to transition the chemical industry towards a sustainable model while ensuring that the policies create jobs, bolster the economy, and fortify energy security, thereby propelling us toward the vision of a decarbonized future.

Electrification must reliably supply energy, mitigating short-term risks such as blackouts and long-term risks like oil/gas market and petrochemical shocks due to geopolitical unrest. Moreover, this must be achieved while ensuring affordability, as the cost - both in terms of average expenditure and price volatility - significantly impacts the feasibility and acceptance of such a transition. Notably, in developing countries, energy equity, which is access combined with affordability and sustainability, becomes paramount.

Sustainability remains a non-negotiable aspect of this transition, with a focus on reducing emissions of NOx, SOx, CO2, and other pollutants. It requires a well-to-gate assessment of lifecycle GHG emissions, from feedstock production through to the point of production, ensuring that upstream and onsite processes align with stringent emission benchmarks. The Clean Air Act is a federal law that regulates air emissions from mobile and stationary sources, and hazardous air pollutants.

This holistic, multi-dimensional approach allows for the understanding of the broad spectrum of factors influencing the decarbonization of the chemical industry. It enables stakeholders to make informed decisions that balance environmental objectives with economic and social considerations, ultimately guiding the industry towards sustainable practices that align with global climate goals. Through such a methodical and inclusive analytical process, the industry can identify optimal pathways for reducing emissions, enhancing energy efficiency, and adopting renewable energy sources, thereby contributing significantly to the global effort to mitigate climate change.

The policy factors capturing this initiative are exemplified in various legislative frameworks, such as Section 45, which incentivizes electricity produced from renewable resources. Crucially, Section 45V is particularly relevant as it provides a credit for the production of clean hydrogen, with the level of credit based on carbon intensity. Such incentives are designed to stimulate the advancement and adoption of low-carbon technologies. This establishes a direct link between the policy frameworks and the operational aspects of decarbonization, ensuring that there is a tangible incentive structure for companies to innovate and reduce their carbon footprint. Nevertheless, does the regulation cover novel technologies?

The integration of these processes with the existing energy grid also poses a significant challenge, requiring robust systems-level analyses to ensure that new electrified processes can be scaled up without compromising the stability of the grid. This involves not only technical considerations but also regulatory frameworks that facilitate such integration, enabling the grid to handle the variable nature of renewable energy sources.

As the narrative of decarbonization unfolds, the deployment of advanced technological solutions like electric heaters, microwave reactors, plasma technology, and electrolysis units is crucial for transforming the chemical industry. These technologies offer reduced emissions, enhanced reaction rates, and energy efficiency, revolutionizing traditional processes.

The evaluation of these technologies through TRLs is essential to gauge their maturity and integration feasibility into industrial settings. This necessitates supportive policies for research and development to upgrade their TRLs. The question is whether the right policies are in place to advocate and upgrade their TRL through research and development.

The practicality of technology implementation involves integrating new technologies into existing systems, considering energy input fluctuations, feedstock variability, and incentives. This requires comprehensive systems-level analyses, including process dynamics simulation, material and energy balance assessment, and environmental impact evaluation. Process intensification efforts might integrate reaction and separation steps, adopt novel catalysts, and explore alternative pathways to increase yield and reduce waste.

In terms of policy implications, the transition towards electrification must be underpinned by robust support mechanisms. These measures help offset the higher initial costs associated with deploying advanced technologies, bridging the gap between current practices and a more sustainable future.

In parallel with technological advancements, research institutions play a crucial role in advancing the state-of-the-art in electrification technologies. Academic and industrial research must continue to explore novel pathways for chemical synthesis and evaluate policies in place.

To maximize the impact of these technologies, grid integration policies and strategies must be developed that can accommodate the variable output from renewable energy sources, including solutions such as smart grids and energy storage systems. Policy effectiveness must be developed and measured via mathematical scenario-based models, roles and responsibilities must be set clearly, the roadmap for energy transition must be delineated, and robust monitoring of progress must be ensured.

Moreover, the decarbonization effort should be inclusive, ensuring that the benefits of a cleaner industry are shared widely. This considers the broader vision of sustainability that encompasses improved public health, environmental justice, creating jobs, and resilience against climate change impacts.

In conclusion, the electrification of the chemical industry represents a challenge but is necessary to further increase profitability and address climate change. It requires an integrated approach that combines quantified systematic methodology for comparing various scenarios considering technological innovation with robust policy support, economic incentives, and societal engagement. By focusing on technology readiness, investing in research and development, and fostering an adaptive regulatory environment, the chemical industry can not only comply with carbon emission regulations but also lead the way toward a more sustainable future for all.

357e Economically Optimal CO_2 Utilization Via Flexible Chemical Production Facilities

Matthew Palys, Prodromos Daoutidis

Hydrogen production via electrolysis powered by renewable energy is emerging as a viable alternative to fossil fuel reforming due to the proliferation and resulting cost reduction of renewable generation and electrolysis technologies as well as production incentives provided by policy such as the Inflation Reduction Act [1]. This "green" or "renewable" hydrogen can be used as a carbon-free fuel for power generation or heating while also being used as a feedstock for other lower carbon intensity chemical production [2]. Toward a broader view on sustainability, carbon dioxide (CO_2) capture and subsequent sequestration or utilization will likely be necessary to meet emissions reduction targets while still producing indispensable commodities such as steel and cement [3]. At the same time, many essential societal functions are enabled by carbon-containing molecules. Prominent examples include urea, methanol, and fuel for long-haul air transportation. Urea accounts for over 50% of global nitrogen fertilizer applied to crops and is made by combining ammonia and CO_2 [4]. Methanol produced from renewable hydrogen can be used to displace conventionally produced methanol as polymer production feedstock and can also act as a carbon-neutral fuel. Renewable methanol can be produced through direct hydrogenation of CO_2 [5]. Aviation is responsible for 3% of global CO_2 emissions and is difficult to electrify using batteries or renewable hydrogen due to energy density requirements. One approach to sustainable aviation fuel (SAF) is to combine hydrogen and CO_2 via either water gas shift followed by Fischer-Tropsch synthesis (WGS-FT) or through a methanol-to-olefins (MTO) pathway [6].

In this work, we optimize the production economics of urea, methanol, and SAF considering only CO_2, water, air, and intermittent renewable energy as feedstocks. We define production facility superstructures for each commodity which include commercially established chemical production processes as well as storage technologies for energy and/or chemical intermediates to facilitate the use of intermittent renewables. We develop a combined optimal design and scheduling framework which minimizes the production cost of each commodity for a given annual capacity of CO_2 to be utilized. This is achieved by optimizing the selection and size of constituent technologies as well as their hourly operating schedules (e.g., production rates and storage inventories) in response to yearlong hourly resolution time-series data for renewable generation. This optimization approach enables systematic consideration of the unique capital investment, energy intensity, and dynamic flexibility (e.g., operating range and ramp rates) of each technology required to produce a given commodity.

We perform a case study for CO_2 utilization for a 50,000 mtCO_2/y to 1,000,000 mtCO_2/y scale range in five regions throughout Minnesota. These regions represent different renewable generation potentials and industries from which CO_2 is emitted. We optimize the production cost of urea, methanol, and SAF while also determining production cost sensitivity to feedstock CO_2 price, capital investment costs for renewable generation and electrolysis, and process flexibility for production of final products or intermediates (e.g., ammonia, WGS syngas). We compare the optimal renewable chemical production costs with market price histories and projections for urea, methanol, and aviation fuel. This comprehensive assessment allows determination of the best CO_2 utilization pathway in the near-term. It also allows for the identification of underlying technological improvements which are critical to the economical and sustainable CO_2 utilization in producing essential commodities.

References

[1] Inflation Reduction Act of 2022, H.R.5376 (2022). https://www.congress.gov/bill/117th-congress/house-bill/5376/text

[2] Clark et al. (2023). Opportunities for green hydrogen production with land-based wind in the United States. *Energy Conversion and Management, 296*, 117595.

[3] Fu et al. (2022). Research progress on CO2 capture and utilization technology. *Journal of CO_2 Utilization, 66*, 102260.

[4] Mao et al. (2024). Green urea production for sustainable agriculture. *Joule*. doi.org/10.1016/j.joule.2024.02.021

[5] Mucci et al. (2023). Cost-optimal Power-to-Methanol: Flexible operation or intermediate storage?. *Journal of Energy Storage, 72*, 108614.

[6] Grahn et al. (2022). Review of electrofuel feasibility—cost and environmental impact. *Progress in Energy, 4*(3), 032010.

357f Integrating CO_2 Electrolyzers in Electrochemical Plants: Heat Integration, Techno-Economic Analysis, and Life Cycle Assessment of the Production of 1-Butene from CO_2

Lukas Spiekermann, Mi Gyoung Lee, Joshua Wicks, Ke Xie, Luca Bosetti, Benedikt Winter, Jan Seiler, Edward H. Sargent, André Bardow

Efforts to mitigate climate change have increased interest in utilizing CO_2 as a sustainable feedstock within the chemical industry [1]. Recent progress in electrochemical CO_2 reduction, facilitated using renewable electricity sources, presents a promising avenue for synthesizing valuable C_{2+} chemicals [2]. While the initial focus has been on the development of electrolyzers, recent research has broadened to include process design and downstream separation techniques [3]. However, the question remains how to integrate feedback from process design back into electrolyzer development.

This work introduces a feedback loop from process design to electrolyzer development. The loop combines process modeling, heat integration, techno-economic analysis, and life cycle assessment.

The resulting workflow is demonstrated for the CO_2 electroreduction to ethylene and subsequent dimerization to 1-butene [4]. From the analysis, we derive development targets for CO_2 electrolyzers. We assess the significance of cell voltage and Faradaic efficiency, highlighting that single-pass conversion minimally impacts overall process feasibility. Moreover, we emphasize the benefits of integrating electrolyzers with downstream and upstream units.

In summary, we combine electrolyzer development and process systems engineering perspectives by bridging the scales from electrolyzers to integrated processes with economic and sustainability objectives. This study illustrates the integration of a process-oriented approach into technology advancement for an electrified chemical industry.

Acknowledgments

This publication was created as part of NCCR Catalysis (grant number 180544), a National Centre of Competence in Research funded by the Swiss National Science Foundation.

References

[1] W. Chung, W. Jeong, J. Lee, J. Kim, K. Roh, and J. H. Lee, "Electrification of CO2 conversion into chemicals and fuels: Gaps and opportunities in process systems engineering", *Comput. Chem. Eng.*, 170 (2023), 108106.

[2] S. Lee, D. Kim, and J. Lee, "Electrocatalytic Production of C3-C4 Compounds by Conversion of CO2 on a Chloride-Induced Bi-Phasic Cu2O-Cu Catalyst", *Angewandte Chemie*, 127 (2015), 14914–14918.

[3] A. Somoza-Tornos, O. J. Guerra, A. M. Crow, W. A. Smith, and B.-M. Hodge, "Process modeling, techno-economic assessment, and life cycle assessment of the electrochemical reduction of CO2: a review", *iScience*, 24 (2021), 102813.

[4] M. G. Lee, X.-Y. Li, A. Ozden, J. Wicks, P. Ou, Y. Li, R. Dorakhan, J. Lee, H. K. Park, J. W.

Yang, B. Chen, J. Abed, R. dos Reis, G. Lee, J. E. Huang, T. Peng, Y.-H. Chin, D. Sinton, and E. H. Sargent, "Selective synthesis of butane from carbon monoxide using cascade electrolysis and thermocatalysis at ambient conditions", *Nat. Catal.*, 6 (2023), 310–318.

357g Cost Efficiency Vs. Energy Utilization: How the Techno-Economic Optimization of Green Ammonia Production Illustrates Novel Influences for an Electrified Chemical Industry

Collin Smith, Laura Torrente Murciano

Decarbonization of the chemical industry through electrification with renewable energy will require the re-optimization of conventional chemical processes considering the novel constraints imposed by renewable energy. Of principal importance is the intermittency of solar and wind energy, which is contrary to the steady supply of fossil-fuels for the chemical industry today. Ammonia production is responsible for ~45% of emissions in primary chemicals production,[1] making its electrification to *green* ammonia production essential to decarbonizing the chemical industry. Further, the well-known Haber-Bosch process for producing ammonia is customarily operated at steady-state, thereby challenging it integration with intermittent solar and wind energy.[2] Therefore, the production of green ammonia using hydrogen generated by water electrolysis, nitrogen separated from air and power from solar/wind energy is a case study which exemplifies the novel constraints faced when electrifying chemical processes.

In this work, by optimizing the techno-economics of green ammonia production with a widespread range of solar and wind energy profiles across Europe, it is shown that intermittent energy reverses the conventional paradigm which pairs increased energy utilization with decreased cost. Chemical processes have for many years been optimized and integrated to maximize the utilization of energy at steady-state. However, daily/seasonal intermittency in renewable energy supply causes the wasting of energy (i.e. curtailment) to be optimal for economic green ammonia production – sometimes exceeding greater than 50% of energy curtailed. Unlike in the case of curtailed energy on the grid, this curtailment is primarily dictated by limitations in the storage capacity of the process rather than the power capacity, which produces unique cost drivers compared to an electrical grid. To facilitate this optimization, neural networks are trained and implemented, illustrating how AI can promote the incorporation of vast spatial-temporal data associated with renewable solar and wind energy.

To assess measures which simultaneously decrease curtailment and cost, this work further optimizes green ammonia production with a combination of solar and wind energy and adjusting (i.e. ramping) the production level of the Haber-Bosch process. While curtailment generally decreases with both process modification, it is unable to completely remove seasonal energy intermittency in all but the ideal locations. Further, when considering energy intermittency at the time-scale of multiple years rather than seasonal intermittency, it is shown that the optimal curtailment and capacity deviates significantly from that predicted from seasonal intermittency due to the unique cost drivers for optimizing constant chemical production with intermittent energy supply. Therefore, this work demonstrates the novel influences which emerge at the process and system level when electrifying a steady-state chemical process with dynamic solar and wind energy.

1. Sanchez, D.P., Collina, L., and Voswinkel, F. (2023). Chemicals. https://www.iea.org/energy-system/industry/chemicals.
2. Smith, C., and Torrente-Murciano, L. (2024). The importance of dynamic operation and renewable energy source on the economic feasibility of green ammonia. Joule *8*. 10.1016/j.joule.2023.12.002.

357h Dynamics of Solar and Wind Power Variability and Implications for Chemical and Material Industries

Micah Ziegler, Shannon Hwang, Aliza Khurram, Jessika Trancik

Increasing deployment of solar and wind power technologies combined with further electrification of energy services could significantly reduce reliance on fossil fuels and greenhouse gas emissions. If energy systems become heavily reliant on solar and wind energy, fluctuations in these resources could present new challenges for reliably providing a responsive energy supply, while also introducing new opportunities to lower system costs for industries that can modulate their demand to match power availability. As such, there is considerable interest in designing and developing chemical and material processes that can utilize electricity flexibly.

In this work, we characterize the size and dynamics of fluctuations in solar and wind resources to elucidate how industrial processes might ease the adoption of renewable resources and increase utilization of electricity infrastructure. Using forty years of weather reanalysis data for the coterminous United States, models of renewable energy technologies, empirical time series of electricity demand, and an energy system model, we simulate the hourly operation of possible future energy systems that rely heavily upon solar and wind power and energy storage to meet demand reliably over decades. We quantify the size of renewable power fluctuations and examine their dynamics over time and in space. We identify common characteristics of meteorological patterns that could lead to resource droughts as well as times when potentially excess energy might be generated. We then quantify how these characteristics might change under different scenarios, which include deploying additional energy storage, expanding transmission infrastructure, and relying on alternative power generators; and we examine how system costs might respond. The results provide insights into how to design new flexible chemical and material processes and schedule their operation in order to enable greater utilization of renewable energy resources and lower electricity costs.

185a Electrified Plasma-Swing Reactive Carbon Capture (RCC) Process for CO_2 Upgrading to Useful Platform Molecules

Chae Jeong-Potter, Stefano Dell'Orco, Nico Dwarica, Susan E. Habas, Daniel Ruddy

Carbon capture and utilization is an increasingly important decarbonization strategy for climate change mitigation. Reactive carbon capture (RCC) technologies, in which capture and conversion of CO_2 occur in a single reactor, are particularly energetically and economically attractive by avoiding the need to purify, compress, and transport the captured CO_2. These technologies are enabled by dual function materials (DFM) that have both sorption and catalytic capabilities to capture CO_2 from dilute sources (i.e. air, flue gas) and convert it to useful C1 platform molecules (e.g. CH_4, CO, MeOH). Most RCC processes are thermally driven, and electrification offers the potential for deeper decarbonization, adaptability, and flexibility. In particular, the use of non-thermal plasma (NTP) offers many potential benefits: (1) activation of relatively unreactive molecules, like CO_2, at low temperatures (≤100°C), (2) rapid onset of reactive conditions to mitigate the release of unreacted CO_2, and (3) ease of on/off operation, facilitating a cyclic operation. Herein, we propose and explore a plasma-swing RCC operation, using a Ru+Na/Al_2O_3 DFM as proof-of-concept.

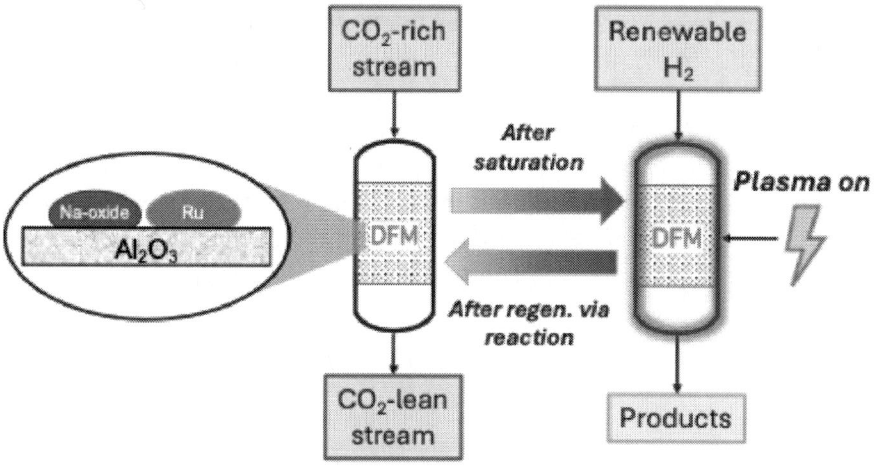

185b Modeling of Electrothermal Reactor for Carbon Capture Using Electric Swing Adsorption

Sairaj Patil, Jay Hyung Lee, Kandis Leslie Abdul-Aziz

The escalating concerns over rising carbon footprints and global temperatures underscore the urgency of minimizing greenhouse gas emissions through methods like post-combustion carbon capture using adsorption processes. Traditional approaches to CO_2 Capture and Utilization (CCU) involve carbonation (capture) on solid sorbents, followed by regeneration and then transportation of CO_2-rich feeds for transformation into high-value chemicals. For desorbing captured CO_2, temperature or pressure swing adsorption processes are commonly used, yet these methods face scalability challenges due to their substantial energy requirements and the necessity for large-scale separation units. In contrast, Electrothermal Integrated CO_2 Capture and Utilization (e-ICCU) offers a more energy-efficient solution by leveraging joule heating (resistive heating of conductors) to heat dual-functional materials (catalyst-sorbent) for CO_2 desorption, then directly converting it to high-value chemicals within a single reactor system. This approach negates the need for the purification and transportation stages associated with captured CO_2. This study introduces a one-dimensional electrothermal reactor model, designed in gPROMS, featuring a Ni-CaO catalyst-sorbent on a SiC mesh (an electrically conductive carrier) for capturing and desorbing CO_2 from flue gas using Joule heating. The model applies a rigorous first-principles process to derive mass-energy balance equations and employs finite element modeling to determine the concentration and temperature distributions across the catalyst-sorbent bed. Utilizing experimental data from a novel laboratory-scale electric reactor, we assess the accuracy of our simulated model, which integrates multiple physical phenomena, specifically diffusion and thermal conduction. This model sheds light on the fundamental processes involved in joule heating of composite materials and CO_2 capture using the electric reactor. It enables the optimization of variables such as joule heating input, reactor design, and sorbent-catalyst loading, aiming to enhance energy efficiency and maximize conversion yields. The future goal of the model aims to integrate a reaction scheme for the catalyst-sorbent using joule heating to convert captured CO_2 to syngas.

185c Integration of renewable energy and reversible solid oxide cells towards decarbonizing the secondary aluminium production and urban systems

Daniel Florez Orrego, Dareen Dardor, Meire Ellen Gorete Ribeiro Domingos, Reginald Germanier, François Maréchal

Abstract

Secondary aluminium is responsible for 2% of industrial emissions, as it depends on natural gas [1]. Efforts are sought to reduce fossil fuel dependency increasing renewable energy (RE) share, using advanced energy systems and untaping savings potentials, like waste heat from furnaces stacks or casting water [2]. A combination of technologies including optimal storage, oxycombustion, carbon capture, power-to-gas (PtG), biomass conversion and electrification are devised to improve process integration and resiliency [3, 4]. In this work, a scenario of waste heat valorization and RE integration is presented, in which waste heat recovery from aluminium plant is used to reduce energy input for scrap metal preheating [2, 5] and space heating purposes [6]. A high temperature reversible solid oxide cell (rSOC) produces power or renewable hydrogen, which can be combined with biogenic CO_2 to produce synthetic natural gas to replace fossil fuel in the aluminum plant [7]. Electricity generated is used in electrical furnaces, rolling processes, and heat pumping systems of district heating network (DNH). Tail gas from rSOC partly supplies heat to melter furnace. These approaches showed promising potential to attain negative emissions and add flexibility to energy and carbon management (Figure 1) [8].

In the aluminium plant, pure aluminium is preheated (250°C) and melted together with scrap aluminium (660°C). Direct chill casting (DCC) solidifies aluminium into ingots producing hot water (50 °C). In the advanced scenario, casting heat is integrated into a CO_2 DHN (~50 bar, 14°C) composed of a closed loop that gathers waste heat from the aluminium plant, biomass conversion and PtG systems and supply space heating or domestic hot water to a nearby urban system. In rolling plant, ingots are treated in pusher furnaces (550°C), hot and cold rolled, and annealed in a continuous line (ACL). In the gasifier (900°C), biomass is converted into syngas using steam [9], which is compressed and purified to produce hydrogen or synthetic natural gas. Apart from CHP production, waste heat is used to generate steam to feed a solid oxide electrolyzer (SOEC, 750°C, thermoneutral, 1.29 V) that produces hydrogen and oxygen [10]. PtG is expected to play a role in harvesting renewable electricity into storable chemicals and fuels [11]. rSOC also works as fuel cell (SOFC, 750°C) consuming hydrogen and oxygen (or air) to produce electricity [12], used to drive other heating furnaces and appliances in aluminium plant, as well as DHN heat pumps [13]. Thus, the number of decarbonization options increases, requiring a systemic analysis to evaluate the most suitable and carbon neutral technologies to supply the energy requirements. The systems are modeled via Aspen ® software or Coolprop, ensuring constraints defined by the minimum energy requirements. Table 1 compares energy consumption remarks of the conventional scenario (i.e. natural gas-fired furnaces and biomass boiler-based DHN) and the integrated solution proposed to decarbonize the aluminium plant and the city demands. A MILP problem is conducted using OSMOSE [14] to maximize waste heat recovery and define favorable configurations, considering the temporal variation of the prices and energy demands.

The integrated scenario (Figure 1) consumes 27% more energy than the conventional scenario, because biomass is converted internally. Yet, conventional scenario is dependent on fossil gas, subject to volatile market conditions and increasing surveillance, due to direct and indirect atmospheric emissions. It is worth noting an opportunity to increase the share of RE in aluminium industry and urban systems by importing nine times more electricity from the grid. In rSOC, the power consumption during the electrolysis mode averages 57 MW, whereas fuel cell power generation is 15 MW. rSOC together with storage systems help managing CO_2 and SNG flows. CO_2 produced and captured (see Table 2) can be later used to generate SNG that is stored in summer season and consumed later in winter season. By harnessing electricity in summer, total biomass energy consumption drops by 25% of the total amount consumed in inefficient biomass boilers. Table 2 summarizes the CO_2 emissions balance including direct and indirect emissions for both studied scenarios. Net CO_2 emissions of the conventional scenario can be reduced more than five times if an integrated carbon management and sequestration system is employed. Captured and injected biogenic CO_2 allows offsetting indirect emissions associated to the supply chains of electricity and biomass.

Figure 2 illustrates seasonal storage profiles for SNG and CO_2 gases. Loading of SNG tanks occurs during summer season when surplus electricity from RE resources is available. It renders more attractive to harvest electricity and store it in the form of fuel to be used later in the colder season. Biomass resource harvested during the summer season can be used to produce syngas or synthetic natural gas consumed in the SOFC. The CO_2 gas is used to refill the storage tanks that store CO_2 until is reacted in a methanation section to produce SNG in summer season. The benefit to store RE in the form of gas is twofold: biomass consumption is drastically reduced and biogenic CO_2 produced is used to offset the CO_2 losses. Annualized investment and operational costs calculated for the conventional and integrated systems, are summarized in Table 3. Total cost of advanced technologies is 28% higher than in the conventional route, due to twenty times higher annualized investment. Yet, the annualized operational expenditure of the conventional solution is higher (11%), due to cheaper biomass resources and use of electricity during the summer season to produce and store fuel using biogenic CO_2.

References

[1] IAI. Aluminium Sector GHG Pathways to 2050, At: https://international-aluminium.org/resource/aluminium-sector-greenhouse-gas-pathways-to-2050-2021 (25 March 2023).

[2] Kumar RA, et al. A Method for Recovering the Waste Heat to Achieve Overall Energy Conservation in Aluminum Casting Industries. Recent Advances in Energy Technologies. Singapore: Springer, pp. 331–342.

[3] Yu M, et al. Waste Heat Recovery from Aluminum Production. In Energy Technology 2018. Springer International, pp. 165–178.

[4] Clos, D, et al. Enabling Efficient Heat Recovery from Aluminium Pot Gas. In Light Metals 2017. Springer International Publishing, pp. 783–791.

[5] Arink T. Metal Scrap Preheating using Flue Gas Waste Heat. Energy Procedia 2017; 105: 4788–4795.

[6] Stijepovic MZ, et al, Optimal waste heat recovery and reuse in industrial zones. Energy, 36, 7, pp 4019-4031.

[7] Borge-Diez, D, et al. Analysis of Power to Gas Technologies for Energy Intensive Industries in European Union. *Energies* 2023; 16: 538.

[8] Florez-Orrego D. et al, A systemic study for enhanced waste heat recovery and renewable energy integration towards decarbonizing the aluminium industry. 36th ECOS 2023, June 25-30, Canaria, Spain.

[9] Flórez-Orrego D, et al. Comparative exergy and economic assessment of fossil and biomass-based routes for ammonia production. EnConvMan 2019; 194: 22–36.

[10] Fan L et al. Advances on methane reforming in solid oxide fuel cells. Renewable and Sustainable Energy Reviews 2022; 166: 112646.

[11] Domingos M, et al. Sustainable Methanol Production Using a Multi-Time Energy Integration and CO2 Management Approach as an Energy Carrier for Synthetic Natural Gas Synthesis. AIChE Annual Meeting. Orlando USA: AIChE, 2023.

[12] Salomone, F, et al. Techno-economic modelling of a Power-to-Gas system based on SOEC and CO2 methanation in a RES-based electric grid. Chemical Engineering Journal, 377 , 2019, pp 120233.

[13] Jensen SH, et al. Large-scale electricity storage utilizing reversible solid oxide cells combined with underground storage of CO2 and CH 4. Energy&EnvironmentalSci, 2015; 8: 2471–2479.

[14] Flórez-Orrego D, Ribeiro Domingos M., Maréchal F. A systematic framework for the multi-time integration of industrial complexes and urban systems, 7th CPOTE 2022, Warsaw, Poland, 20-23 September, 2022.

438a Calcium Mediated Electrochemical Synthesis of Ammonia

Ishita Goyal, Meenesh Singh

The synthesis of green ammonia stands as a pivotal advancement towards fostering a sustainable and secure future, with applications ranging from agricultural fertilization to energy storage. To harness ammonia sustainably, innovative processes are imperative to mitigate the substantial carbon footprint and energy consumption inherent in conventional methods like the Haber-Bosch process. Electrochemical synthesis of ammonia emerges as a promising avenue, leveraging renewable resources for electricity consumption. Li mediated NH3 synthesis is a promising approach to make NH3 at ambient conditions and it has been widely investigated. This work explores the electrochemical calcium-mediated synthesis of ammonia (Ca-MAS) as an alternative approach. The process initiates with the electrodeposition of calcium metal on a Ni foam substrate, followed by nitridation, culminating in ammonia formation. While lithium-mediated synthesis has been extensively investigated, its reactivity poses safety concerns. Calcium, with a lower equilibrium potential for deposition and abundant availability compared to lithium, offers promising performance and economic viability. A comprehensive study examines various electrolyte compositions, including base electrolytes such as tetrahydrofuran (THF), propylene carbonate, acetonitrile, dimethoxy ethane, diethylene glycol dimethyl ether, and dimethyl sulfoxide (DMSO), considering factors such as volatility, conductivity, and stability. The effects of these compositions on performance metrics such as faradaic efficiency (FE) and current densities are analyzed. This work aims to achieve comparable yields and current efficiencies to lithium-mediated synthesis, thereby reducing associated costs and enhancing the environmental sustainability of ammonia production.

438b The Electrochemical Partial Oxidation of Methane to Methanol on MnOx.

Ahmed Almunaifi, Carlos Morales-Guio

The utilization of greenhouse gases to produce energy and products is of great interest. The direct electrochemical partial oxidation of methane to methanol is of particular interest in the oil and gas industry. Previous efforts of our group have been employed to study the effect of a range of electrodeposited transition metal oxide films on a cylindrical electrode through a series of experiments and DFT calculations. The work proved certain transition metal oxides to be active for the electrochemical transformation of methane to methanol with high selectivity. The previous work thoroughly studied cobalt (oxy)hydroxide. In this work, we focus our analysis on the performance of manganese (oxy) hydroxide as a catalyst for the partial oxidation of methane. Namely, we test the effect of the applied potential and the rotation speed on the selectivity of methanol production. The results show that at intermediate oxidative potentials and high rotational speeds both methanol and acetate are produced; however, MnOx catalyzes the partial oxidation reaction with high selectivity towards acetate production.

438d Enhanced Electrochemical Hydrodeoxygenation of Bio-Oil Model Compounds Using Bimetallic Platinum Catalysts

Jeffrey Page, Amol Pophali, Tae Jin Kim, Juan A. Lopez-Ruiz, Stoyan Bliznakov, Julia A. Valla

The production of renewable fuels has become necessary as fossil fuels continue to deplete past the point of no return.[1] Bio-oils derived from the fast pyrolysis of biomass represent a promising source of renewable and sustainable, carbon-neutral chemicals and fuels. Pyrolysis oils, however, struggle from low yields and increased heteroatom content.[2] To solve this problem a second upgrading step is needed which increases the fuel's hydrogen content and removes heteroatoms. Conventional oil upgrading has been done through thermochemical processes using high temperatures (>300 °C) and high-pressure hydrogen (10-20MPa).[3] Electrochemical hydrodeoxygenation (EC-HDO) is a promising method for upgrading biomass derived oxygenates into biofuels at near ambient conditions without the need for external hydrogen (H2).3 Though initial results on electrocatalytic upgrading have been promising few researchers report appreciable yields of fully deoxygenated products and the efficiency remains low at high current densities.[3]

This work presents the upgrading of phenol and guaiacol as bio-oil model compounds. EC-HDO of phenol and guaiacol were investigated in a custom electrochemical cell (H-Cell), wherein the anode and cathode are separated by a Nafion® 117 membrane. Preliminary results on the EC-HDO of phenol using a lab synthesized PtRu-C catalyst showed that after 4 hours of electrolysis at -0.46 V vs Ag/AgCl cyclohexane selectivity reaches 31%. Further EC-HDO experiments were performed using lab synthesized Pt and Ru catalysts to investigate the role of active metal on cyclohexane selectivity. These results (figure 1) showed that the lab synthesized PtRu catalyst produced the highest cyclohexane selectivities (31%), highest conversion (100%) and efficiencies (35%). Additional experiments on EC-HDO of guaiacol showed that using the lab synthesized PtRu-C catalyst resulted in a cyclohexane selectivity of 34%. Our preliminary results indicate that the addition of metals to Pt-C catalysts can increase phenol EC-HDO performance and cyclohexane selectivity. Our current work is focused on the development of new bimetallic catalysts for the enhancement of cyclohexane productions such as PtCo, PtNi, and PtCu. Additionally, experimental results will be presented using guaiacol as a model compound to show how the bimetallic catalysts affect the EC-HDO of various functional groups.

References

(1) Zhang, X.; Lei, H.; Chen, S.; Wu, J. Catalytic Co-Pyrolysis of Lignocellulosic Biomass with Polymers: A Critical Review. *Green Chemistry*. 2016. https://doi.org/10.1039/c6gc00911e.

(2) Zhang, B.; Zhong, Z.; Min, M.; Ding, K.; Xie, Q.; Ruan, R. Catalytic Fast Co-Pyrolysis of Biomass and Food Waste to Produce Aromatics: Analytical Py-GC/MS Study. *Bioresour. Technol.* **2015**, *189*. https://doi.org/10.1016/j.biortech.2015.03.092.

(3) Chen, G.; Liang, L.; Li, N.; Lu, X.; Yan, B.; Cheng, Z. Upgrading of Bio-Oil Model Compounds and Bio-Crude into Biofuel by Electrocatalysis: A Review. *ChemSusChem*. 2021. https://doi.org/10.1002/cssc.202002063.

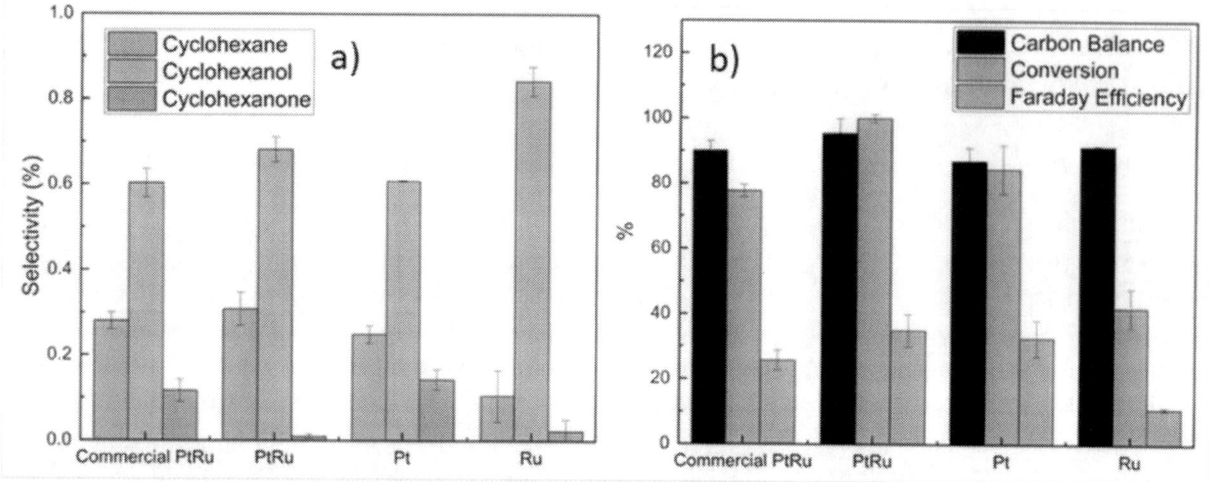

Figure 1. a) product selectivities for the EC-HDO of phenol. b) carbon balance, conversion, and faraday efficiency for phenol ECH with different catalysts. Reaction conditions: 55 mA cm^{-2}, 50 mL min^{-1} N_2, 60 °C, 0.2 M perchloric acid, 0.2 mg cm^{-2} catalyst loading.

438e Natural Gas to Hydrogen Via a Process Intensified Cold Atmospheric Plasma-Based Reformer

Shayan Niknezhad, Efstratios Pistikopoulos

This research underscores the paradigm that challenges in reforming fossil fuels into cleaner energy sources lie not within the feedstocks themselves but in the processes utilized. By developing a 'clean' chemical process via Cold atmospheric plasma-based reforming, we aim to fundamentally mitigate CO2 emissions, steering away from traditional combustion-based methods that are less efficient and environmentally harmful. Electrifying the current system has a high potential to reduce emissions and increase the efficiency of chemical processes. Technologies, including electric heating, RF, microwaves, plasma reactors, and electric furnaces, are some options that can be researched and developed for the near future.

The current research aims to develop plasma technology for chemical processes at low temperatures. Low-temperature plasma (LTP) operates outside thermal equilibrium, exhibiting higher electron and vibrational temperatures than their rotational and translational counterparts in molecules. This characteristic makes LTP especially suitable for chemical processes, initiating reactions at low temperatures and power through electron collisions that produce highly reactive species. The vibrational excitation, explained by the ladder-climbing mechanism leading to molecular dissociation and the formation of radicals via electron impacts, are foundational to the chemical processes in LTP.

In LTP, radicals form first by electron impact reactions, and once radicals are generated, propagation and recombination reactions occur to obtain final products. These reaction pathways are generally agreed upon for DC macro and micro-plasma discharges. Vibrationally excited molecules can significantly accelerate chemical reactions by effectively decreasing activation energy. The appeal of LTP extends to chemically reacting atmospheric pressure plasmas, including cold atmospheric pressure plasmas (CAP) like microplasmas, coronas, and dielectric barrier discharges utilized in ionizers. These systems operate efficiently at low temperatures and atmospheric pressure, reducing costs and system complexity. However, the kinetic effects of excited species on chemical processes in non-equilibrium plasma remain partially understood, presenting a significant research opportunity.

To some degree, all gases have some small fraction of ionized particles and free electrons for two main reasons. The first of which is their finite temperature, which leads to a small portion of particles having adequate internal energy to be ionized, as expressed by the Saha equation. The second reason is due to the cosmic sources of ionization radiation. Plasmas are created when there are a sufficient number of charged particles for collective behavior in the medium. The collective behavior is called the plasma approximation and is valid when the electron density (or charged particles) within the sphere of influence of the particle is large enough.

The fundamental inquiry of mitigating CO2 emissions led to the conceptual realization that reducing CO2 could be achieved by circumventing its production. Traditional methods of reducing emissions have predominantly centered on post-production capture and sequestration, but this approach is often costly and not entirely effective. The core concept centers around the transformative potential of plasma-based technology, offering a novel approach to reforming hydrocarbons into valuable products while circumventing the release of direct CO2 emissions. The core concept centers around the transformative potential of cold atmospheric plasma-based technology, offering a novel approach to reforming hydrocarbons into valuable products while circumventing the release of direct CO2 emissions.

Methane is chosen as the feedstock to produce hydrogen for further research on the experimentally developed CAP reformer. The backdrop against which this innovation emerges is the critical role of hydrogen as a raw material in various industrial processes, including fertilizer production, food processing, metal treatment, and petroleum refining. The dominant method for hydrogen production, steam methane reforming, currently accounts for a staggering 90% of hydrogen production in the USA. However, this process comes with significant drawbacks. Annually, it results in the emission of 800 million tonnes of CO_2 in the USA alone, equivalent to the combined emissions of the UK and Indonesia. Additionally, high temperatures (around 800°C) are required to reform natural gas into hydrogen, which is both energy-intensive and environmentally harmful.

In this reforming system, the feedstock (natural gas) passes through a nonthermal plasma-based reformer and reforms to hydrogen at ambient temperature and pressure without the presence of O_2. Then, the produced gases and hydrogen exit the reformer toward a membrane separation. The produced H_2 separates from other produced light hydrocarbons and methane. The retentate gasses from the membrane return to the beginning of the process and mix with natural gas feedstock to reform to H_2.

Basically, the system consists of three main parts: (a) the nanoseconds plasma-based reformer, (b) the membrane and separation part, and (c) the compressors for recirculating gas through the system. The nanosecond plasma-based reformer consists of two main components: (1) plasma reactor and (2) electric power supply. When a high voltage pulse is applied across electrons, neutrons and ions will be accelerated. As enough energy is implied, electrons gain enough energy to break through and flow between two electrodes. While electrons move from anode to cathode, electrons and ions, collide with other neutral molecules in the reactor, creating more electrons and ions.

Pulse parameters include two main factors: the input voltage and the pulse frequency repetition. Consequently, the power input and specific energy input can be calculated. As a result, increasing applied voltage directly increases H_2 production. Observations show that the maximum H_2 production reaches 36.6% of molar concentration at 30 KV. The SEI is a critical factor in evaluating the performance of plasma-based reforming. SEI is the input energy per unit of input mass of methane and is linearly related to the input energy. Changing PRF (pulse reputation frequency) increases the SEI at a higher rate in comparison to the input voltage. Hence, in order to reduce the input energy, a lower frequency is preferred, while enough pulses are required to reform natural gas.

SEI depends on the input power and flow-in rate. Generally, increasing power enhanced CH_4 conversion and H_2 production. The maximum hydrogen production is achieved at 30 W.

At minimum input power (10 W), 23% of methane molar flow was reformed, and at the maximum power (30 W), 60.1% of methane was converted. The H_2 production at the minimum and maximum input power is 14.9% and 34 %, respectively. However, the optimum condition is defined as the highest H_2 yield (kg/kJ). As experimental studies show, at higher flow rates, more pulses are required for reforming. In general, at higher flow rates, the produced hydrogen is higher. However, techno-economic analyses are required to find the most profitable operation condition.

Four scenarios are assumed to study the effect of flow rate through each reactor and the total flow of the plant on Net H_2 revenue. Cases 1 and 3 studied the 1.1 l/min flow through each reactor and the total flow rate of 50000 and 100000 m3/hr. Cases 2 and 4 explain the flow rate at 2.2 l/min at each reactor and the entire plant flow of 50000 and 100000 m3/hr. As concluded, increasing the flow rate through each discharge and system flow rate increases the Net H_2 revenue. The highest achieved Net H_2 revenue is at 2.2 l/min flow through each reactor with a total plant flow of 100000 m3/hr. Sensitivity analyses have been done to study the effect of flow rate and find the upper and lower boundaries of operation.

Aspen Plus is utilized to simulate both input and output streams by solving the mass and energy balances. The base case scenario is a plant feedstock of 100000 m3/hr. A detailed methodology is used to perform a techno-economic analysis on the nanosecond plasma-based hydrogen plant. The results are compared with electrolysis and SMR plants with and without a carbon-capturing system. As a result, the natural gas demand for the plasma reformer plant is 53.72% lower than the SMR plant. By comparing the energy demand for plasma-based reformers with electrolysis technology, it can be concluded that plasma-based reforming requires 87.57% less energy. Based on the unit of energy, the H_2 production yield for the plasma-based plant is 44.42 kg/MW, and for the electrolysis-based plant is 20 kg/MW.

Regarding the environmental impact of the hydrogen production pathways, direct emissions are exclusively higher in the SMR processes. The CAP plasma-based reformer emits zero CO2 directly in the process. A typical SMR plant without CCS releases 9.2 kg CO2 for producing one kilogram of H2. This carbon dioxide can be reduced to 1 kg CO2/ kg H2 by hiring CCS systems. The CAP plasma-based technology produces hydrogen for 2.89 $/kg H2. The Levelized cost of H2 production for electrolysis, SMRs, and SMRs with CCS are 4.51, 1.53, and 1.95 $/kg, respectively.

In summary, a novel nonthermal plasma-based reformer has been created to reform natural gas to hydrogen. A separation system is designed to separate hydrogen from other co-products. The CAP plasma-based reformer was created and went under investigation to find its optimum operating condition. The optimal operating condition for the plasma-based reformer is achieved by experimenting and studying the effect of changing reactors' flow rate, input energy, and pulse frequency repetition. The optimum operating condition is defined as the maximum accomplished H2 yield (kg/kJ). As a result, the process can compete with benchmarked technologies with zero direct CO2 emission and no water consumption.

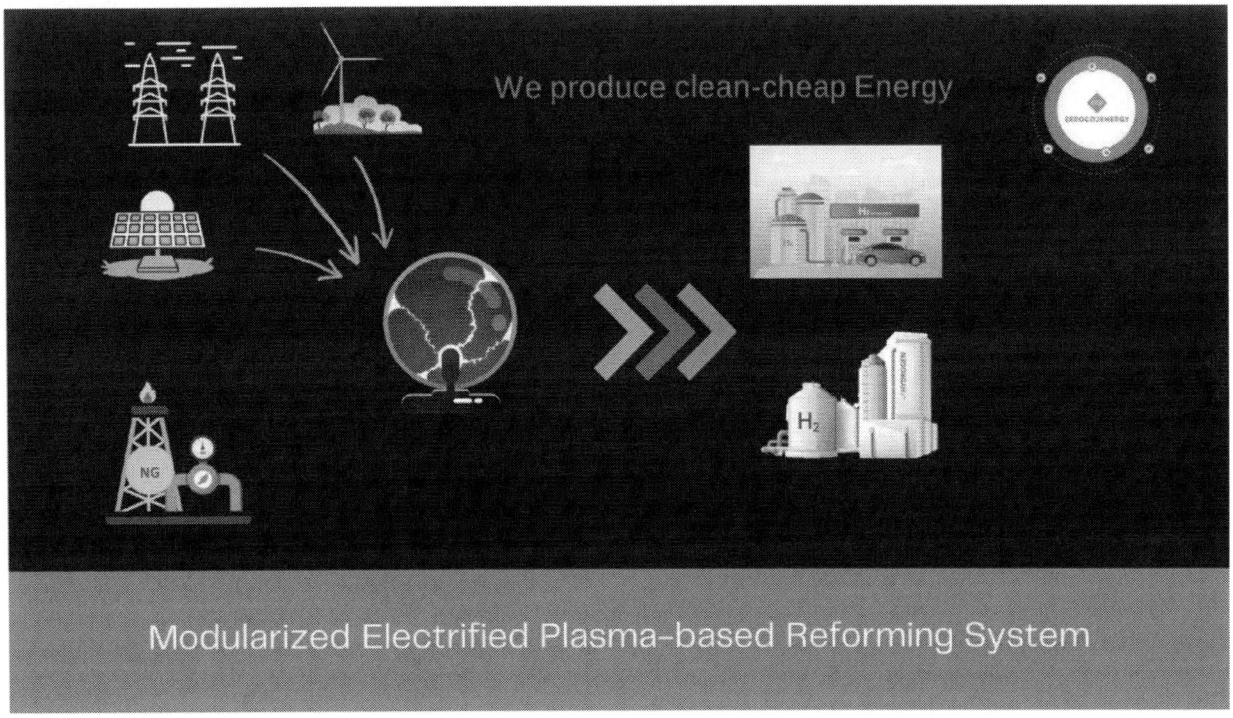

438f Synergizing CO_2 Capture and Electrochemical Conversion: Tailored Oxygen Vacancies and Amine-Based Electrolytes for Enhanced CO Production

Abdul Hakeem Anwer, Assem Mohamed, Maroua Sadaoui, Abdelbaki Benamor

Advancing electrochemical CO_2 reduction reactions (ECO_2RR) is pivotal in the urgent task of mitigating CO_2 emissions. Catalyst development focuses on enhancing CO_2 adsorption and activation, yet strategies addressing product desorption are underexplored. Drawing from the Sabatier principle, our study aims to enhance ECO_2RR performance, achieving an impressive 85% faradaic efficiency for CO production by optimizing product desorption. Through tailored manipulation of the electronic environment, utilizing oxygen vacancies (O_{vac}) in Ni-doped $SrTiO_3$, we lowered the energy barrier for product desorption. Introducing Ni^{2+} into the $SrTiO_3$ lattice promotes O_{vac} formation, altering the local electronic landscape. This doping-segregation method provides a unique avenue for controlling active metal cluster size and oxide support properties, promising diverse chemical reaction applications. Additionally, we present a novel methodology integrating CO_2 capture and conversion, utilizing CO_2 capture media directly as the electrolyte for electrochemical CO_2 conversion, further advancing the feasibility and sustainability of CO_2 utilization technologies.

438g Kinetic Modeling Investigation of the Chemical Opportunities of Ethane Reforming in a Nonthermal Plasma, an Electrified Approach to Natural Gas Valorization

Denver Haycock, Russell J. Clarke, Jason Hicks, William Schneider

Finding electrified routes for the valorization of ethane, usually the second most abundant hydrocarbon in natural gas, has significant environmental benefits and economic incentive. Nonthermal plasma is presented as a possible technology due to its electric-driven nature, mild operating conditions, and demonstrated ability to dehydrogenate ethane. Nonthermal plasmas are generated by applying a strong electric field to a gas in such a way that it becomes partially ionized (~1% ionization) and electrically conductive. This creates a highly reactive environment that is in a non-equilibrium state, with radical, excited, and neutral species near ambient temperatures and free electrons in excess of 10,000 K. Due to the complexities of nonthermal plasma chemistry, most studies using nonthermal plasmas with ethane have focused on demonstrating ethane conversion within a limited range of reaction and plasma conditions. Therefore, we present an investigation of the chemical opportunities within an expanded range of conditions through the use of time-dependent kinetic modeling. We apply a kinetic modeling software (ZDPlasKin) developed for the unique combination of thermally- and electron-excitation-driven chemistry present in nonthermal plasmas. Using a model verified by base-case experimental measurements has allowed us to probe conditions that are often difficult to isolate in an experimental setting. The results of this modeling investigation reveal a landscape of parameter space that expands our understanding of the chemical opportunities of ethane reforming in a nonthermal plasma. We track the product yield and ethane conversion as major outputs of our model, and explore the impact of various inputs including time in plasma, reduced electric field, electron density, and starting gas composition. Within the multi-dimensional results of our modeling, we find opportunities to optimize ethylene and acetylene yields, minimize methane or other less-valuable products yields, maximize ethane conversion, and optimize energy use. For example, limiting the time in plasma to a few micro seconds and increasing the reduced electric field to over 200 Td provides a maximum ethylene yield while keeping methane yield relatively low. A limited technoeconomic analysis based on the possible outcomes of ethane reforming in our model is presented. Our results also highlight opportunities to implement catalyst chemistry to potentially enhance product selectivities.

438h Simple, Cost-Effective Synthesis Method of Electrocatalysts for Electrified Hydrocarbon Production

Ian Street, Sriram Mosali, John Regalbuto, William Mustain

The decarbonization of industrial chemical processes is imperative if humans are going to continue to improve their overall quality of life while simultaneously reducing their impact on the environment. Electrification coupled with a renewable energy supply can facilitate lower emissions through several pathways, though two in particular have been gaining wide popularity in recent years: 1) replacing thermal heating (e.g. burning fuels) with alternative methods (resistive, microwave, etc.); and 2) by replacing thermochemical conversion with electrochemical conversion, where electrical energy can be used directly to drive new chemical reactions. One attractive option that both opens new mechanistic pathways and removes CO_2 from the atmosphere is the electrochemical reduction of carbon dioxide (CO2RR) to hydrocarbons [1].

CO2RR occurs through a complex reaction mechanism that consists of multiple proton and electron transfer steps along a variety of reaction paths to produce mostly C1 and C2 hydrocarbons (carbon monoxide, formate, methane, ethylene, etc.). Because most bulk metals (e.g., Au, Ag, Zn, Pd, Pb, In, and Sn) have lower affinity for important CO2RR intermediates, namely carbon monoxide, the resulting transformations tend to yield some ratio of carbon monoxide and formate only. In contrast, copper catalysts have gained increasing attention due to their ability to more optimally bind CO2RR intermediates to form a litany of products, but competition with the HER and poor selectivity between the multitude of reaction pathways have limited the proliferation of the technology [2].

Single atom catalysts (SACs) have an inherent advantage over bulk metal and metal nanoparticle catalysts because they maximize metal active site utilization for minimal metal loadings. Along with maximizing dispersion, SACs have also been shown to increase the activity of metals for certain reactions, including CO2RR. Lighter transition metals like Fe, Co, and Ni tend to be active mainly for HER and poisoned by carbon monoxide. However, SACs of these metals change their properties and open the door for cheaper and more abundant catalyst materials [2]. There have been numerous techniques developed to synthesize single atom catalysts, but many use high temperatures and specialized chemicals that make them expensive especially when applied to industrial scale [3].

This presentation will report the synthesis and characterization of Cu and other transition metal SACs on high surface area carbon supports for CO2RR. Results from a wide array of physical characterization techniques will be presented, including powder Xray diffraction and STEM imaging, showing the presence of the SACs. The activity and product selectivity for the SACs will be reported in both batch and flow-through electrochemical reactors and linked to the catalyst structure and Cu dispersion. For all catalysts, the product profile, faradaic efficiency, selectivity, and carbon efficiencies will be reported – determined through a combination of integrated gas chromatograph-mass spectrometry and nuclear magnetic resonance.

References:

[1] Xia, R.; Overa, S.; Jiao, F. Emerging Electrochemical Processes to Decarbonize the Chemical Industry. JACS Au 2022, 2 (5), 1054–1070. https://doi.org/10.1021/jacsau.2c00138.

[2] Johnson, D.; Qiao, Z.; Djire, A. Progress and Challenges of Carbon Dioxide Reduction Reaction on Transition Metal Based Electrocatalysts. ACS Applied Energy Materials 2021, 4 (9), 8661–8684.

https://doi.org/10.1021/acsaem.1c01624.

[3] Ji, S.; Chen, Y.; Wang, X.; Zhang, Z.; Wang, D.; Li, Y. Chemical Synthesis of Single Atomic Site Catalysts. Chemical Reviews 2020, 120 (21), 11900–11955. https://doi.org/10.1021/acs.chemrev.9b00818.

616a Microwave-Induced Synthesis of Green Ammonia As a Hydrogen Carrier: A Complete Cycle from Synthesis to Decomposition

Alazar Araia, Yuxin Wang, Brandon Robinson, Jianli Hu

The global landscape is transitioning at unprecedented speed towards decarbonized energy sources. Hydrogen plays a pivotal role in this fast-paced transition and has gained significant attention as an alternative energy source. However, hydrogen is currently produced through a carbon-intensive process that eventually contributes to substantial greenhouse gas emissions, in addition to the low volumetric density and incompressibility making it difficult to store and transport. It is imminent that there is a need to develop novel cost and energy-efficient chemical processes and technologies to produce, store, and transport hydrogen energy produced from renewable energy sources such as wind and solar. The viable option is to develop scalable technologies for converting renewable electrical or thermal energy into chemical energy contained in energy-dense Carbon-Neutral Liquid Fuels (CNLF). These carbon-neutral fuels can be stored, transported, and later converted into the desired renewable hydrogen energy, facilitating the transition into decarbonized energy sources.

Ammonia has attracted significant attention recently due to its unique capability to store and transport renewable hydrogen energy without emitting carbon dioxide. According to the U.S. Department of Energy (DOE), ammonia is categorized as a carbo-neutral liquid fuel when produced from renewable energy sources. On a large industrial scale, ammonia is produced through the Haber-Bosch (HB) process that operates at high temperatures and pressure. The scale-down of the HB process to meet renewable energy requirements will significantly increase the cost of ammonia production. As a result, the search for innovative modular technology adaptable to renewable energy sources is crucial. Microwave technology is adept at producing renewable hydrogen energy due to its unique compatibility with intermittent renewable energy sources. This unique ammonia production using microwave technology is paramount in expanding carbon-neutral hydrogen production while mitigating the current climate crisis.

In our study, we performed a complete cycle of renewable hydrogen production using ammonia as a carrier. Our research study comprises two phases, the first step is focused on ammonia synthesis, while the second step is on ammonia decomposition for hydrogen generation. In this two-phase cycle, the integration of microwave technology and a novel electromagnetic susceptible catalyst that is active and durable for both phases is the core innovation. The employment of microwave technology is unique and promising simultaneously as the heating mechanism is selective and efficient, lowering various reaction conditions such as activation energy, time, temperature, and pressure overcoming the thermodynamic limitations. Developing novel dielectric Cs-Ru/CeO_2 catalysts is also pivotal in achieving green hydrogen under microwave reaction conditions. Ru is chosen as the active metal catalyst due to its superior catalytic activity at lower temperatures than traditional Iron-based catalysts. However, Ru is an expensive rare earth metal and must be supported. CeO_2 is promising support for Ru metal due to its reversible oxidation while Cs is selected as a promoter owing to its high electro-positivity. This novel approach produced ammonia at low temperatures and pressure, while this same catalyst demonstrated efficacy in decomposing ammonia under microwave irradiation.

In our comprehensive study on renewable hydrogen production via ammonia as an energy carrier, ammonia was synthesized at a notably lower temperature of 260 °C, less than 20 bar under microwave irradiation compared to 400-500 °C, 250-300 bars in a traditional thermal fixed-bed reactor over a Cs-Ru/CeO_2 catalyst. In the subsequent phase of ammonia decomposition, the same catalyst was washed-coated onto SiC foam, achieving complete ammonia decomposition at a low temperature of 360 °C, yielding a substantial hydrogen production rate.

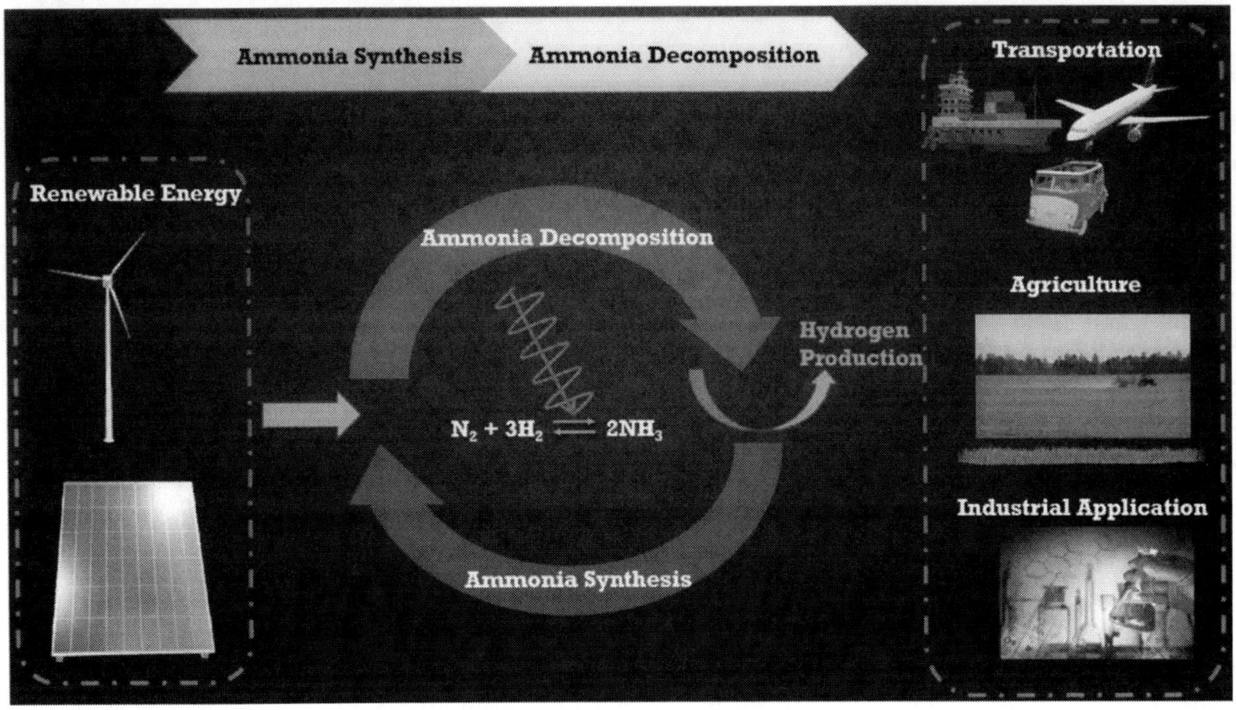

Ammonia Synthesis
Ammonia Decomposition
Transportation
Renewable Energy
Ammonia Decomposition
Agriculture
Hydrogen
Production
$N_2 + 3H_2 \rightleftharpoons 2NH_3$
Industrial Application
Ammonia Synthesis

616b Application of Induction Heating to Endothermic Reactions

Ben Ko, Han Wang, Kian Kwa, Erdem Sasmaz

The application of induction heating to supply heat for chemical reactions has garnered attention due to its potential to electrify chemical processes. Induction heating has the potential to enhance both production and energy efficiency, making it a compelling and environmentally friendly heating method for the chemical industry. Induction heating involves the use of a high-frequency alternating magnetic field on ferromagnetic materials, which then produce heat in response to the magnetic field through hysteresis loss and eddy currents. By placing ferromagnetic susceptors inside the catalytic bed, the catalysts can be internally heated within an induction coil. This improves the temperature distribution in the catalyst bed and provides a prompt heating response for temperature control. Traditionally, catalytic reactions rely on furnace reactors that heat the catalyst bed from the reactor walls, resulting in non-uniform temperature distribution and inefficient dehydrogenation. In contrast, induction heating offers the potential to internally heat the catalyst bed, thereby homogenizing temperature distribution and enhancing catalytic performance.

Endothermic reactions play a critical role in the petrochemical industry, but they are known for being highly energy-intensive and facing challenges such as temperature control, catalyst deactivation, and energy consumption. In this presentation, we will provide an overview of the application of induction heating for endothermic reactions, including ethanol-to-1,3 butadiene, propane dehydrogenation, and ethanol-to-acetone processes. We will demonstrate the impact of induction heating on selectivity and reaction mechanism and discuss the impact of temperature distribution and the rapid compensation of heat loss through induction heating on catalyst behavior.

616c Feasibility of Electrically Heated Wired Reactors for Methane Reforming and Ammonia Decomposition Processes: Decarbonization of Hydrogen Supply Chain

Ram Ratnakar

Continuously increasing global emissions of CO_2 and other greenhouse gases in the atmosphere pose great challenge due to their adverse ecological, health and economical effects on to our society. One of the primary sources of CO2 emissions are chemical and manufacturing industries, where energy is supplied by burning fossil fuels. While several renewable pathways are being on development, enabling these resources for manufacturing industries while improving thermal efficiency are still major gap. In this work, we describe the electrified reactor configurations that can enable the integration of renewable power with manufacturing processes where endothermic heat is required. Hydrogen production is considered as a case example that could include not only the decarbonized heating pathways but also for production of carbon-free fuel. The proposed reactor configurations are shown to have technical feasibility for various endothermic processes including ethane cracking, steam//dry/bi- reforming of methane and ammonia cracking, and can easily be extended for other endothermic processes.

Keywords: Process Intensification; Process Electrification; Net carbon zero, Hydrogen production, Methane reforming, Ammonia cracking

616d Radio Frequency Heating for Catalytic Propane Dehydrogenation

Ankush Rout, Somtochukwu Lambert, Debalina Sengupta, Mark Barteau, Benjamin Wilhite, Micah Green

Here, we demonstrate radio frequency heating of susceptor nanomaterials coupled with conventional catalysts to enable a new class of heterogenous, endothermic catalytic reactors with localized, volumetric heating. Heating in industrial scale catalytic reactors is carried out in fuel-fired furnaces which contributes to CO_2 emissions and requires substantial infrastructure. This work uses the power-to-chemicals route where radio frequency fields (1-200 MHz) are utilized to volumetrically heat up the RF-responsive carbon nanomaterials integrated with the catalyst; by using renewable electricity sources, the greenhouse gas emissions are reduced. In-house built applicators were used for heating the catalyst to very high temperatures (~550 ^oC). We demonstrated the use of one such applicator to drive a highly endothermic propane dehydrogenation reaction on Pt/Alumina catalyst and the RF susceptor is carbon nanotubes. The propane conversion and propylene yield using RF heating was comparable to oven heating at 500 ^oC. After each reaction cycle, the catalyst was successfully regenerated by RF heating. This setup can be used in place of conventional catalytic reactors enabling distributed manufacturing of chemicals.

616f Volumetrically Heated Electrified Chemical Reactor Systems with Characterization As Artificial Electromagnetic Media

Jonathan Fan

Heavy industry is one of the most challenging sectors of the economy to decarbonize due in part to the dominant role of fossil fuels as an energy source for producing high grade heat. To this end, the electrification of thermochemical processes with clean electricity has great potential as both an immediate and long-term solution towards the scalable decarbonization of chemical production. In this talk, I will discuss new paradigms in thermochemical reactor design in which the reactors are volumetrically heated using high frequency magnetic induction. We focus on induction over other heating methods due to its scalability in power and volume. A principal focus is to treat the chemical reactor as an artificial electromagnetic medium that can be tailored and co-designed with the power electronics to enhance heating efficiency and specify the heating profile in a manner that optimizes reaction engineering properties.

I will first discuss metamaterial reactors in which a volumetric open-cell lattice baffle is designed and inductively heated using a helical magnetic coil. The baffle simultaneously serves as a heating susceptor and support for catalyst impregnation. The baffle layout and power electronics are co-designed to ensure the heating profile is volumetric and heating efficiencies are maximized. We first consider a 1.5" diameter reactor containing a volumetric silicon carbide-based foam baffle, which can be modeled as a uniform effective homogeneous medium and which can be heated by megahertz power electronics with over 90% efficiency. We show the reactor can be configured to perform the reverse water gas shift (RWGS) reaction at 600 C and that the reactor operates in ideal plug flow conditions with radially uniform heating profiles. We further show in a scale up analysis that the energy efficiency of the system naturally increases with reactor size and that total system efficiency is only limited by AC-DC power conversion. Finally, we show how the baffle architecture can be tailored to accommodate arbitrary volumetric heating profiles in a manner that further enhances chemical conversion and reduces reactor size.

I will then discuss a new class of resonant metamaterial reactors in which the reactor is a volumetric magnetic resonator. The structure of the reactor is that of a sheet of metal rolled in to a "Swiss roll" geometry. Our resonant metamaterial reactor has multiple features. First, when inductively heated at resonance, the coupling efficiency of the induction system is near unity and is nearly independently of reactor and coil geometry. Second, an individual Swiss roll structure does not get heated at off-resonant frequencies, meaning that Swiss roll structures specified to resonate at different frequencies can be independently addressed and heated via frequency multiplexing. Third, Swiss roll structures get uniformly heated and have exceptionally high surface area, eliminating heating and heat transfer bottlenecks in thermochemical reactions. We design and demonstrate a Swiss roll reactor that performs the RWGS reaction with high (95%+) heating efficiencies and uniform volumetric heating profiles.

Finally, we show that our concepts in the co-design of reactors and power electronics can extend to new reactor modalities outside of fixed bed systems. As an example, we show that molten salt reactors can be heated to very high temperatures (~1000 C) through the direct inductive heating of the salts themselves. In this effort, large signal AC impedance spectroscopy is used to characterize the electromagnetic properties of molten salts, and these properties are used to co-design the power electronics frequency and reactor form factor to enable volumetric heating with coupling efficiencies of over 90%. These reactors can be utilized for methane pyrolysis, and we show that highly efficient magnetic induction is sustained when the reactor is operated as a bubbling bed.

616h Modeling CO_2-Free Hydrogen Production Via Microwave-Driven Methane Pyrolysis in a Fluidized Bed of Carbon Particles

Mehdi Salakhi, Murray Thomson

Hydrogen is expected to play a crucial role in the transition to a net-zero emissions future by 2050. It is a carbon-free energy carrier that is widely used in many industrial applications such as oil refineries, steelmaking, and chemical production. Therefore, hydrogen is a promising vector for supplying global energy demand, and its demand is projected to double by 2030. Currently, only 0.5 Mtonnes/year is provided by clean technologies of hydrogen production, while the global demand for hydrogen is 70 Mtonnes/year. Thus, it is imperative to explore CO_2-free technologies for hydrogen production. Despite the increasing interest in hydrogen via electrolysis, this method is highly energy-intensive and not scalable enough to meet the demand. Alternatively, methane pyrolysis can produce hydrogen without direct CO_2 emissions and with modest energy inputs, making it a potential stepping stone from fossil fuels to renewable energies. Abundant resources of natural gas and the need for CO_2-free hydrogen production methods have made methane pyrolysis a promising area of research and technological development.

Provided by renewable electricity, microwave energy is a sustainable and efficient method of providing energy for methane pyrolysis. In comparison to electrolysis, the use of microwaves significantly reduces electricity consumption by 80%. The current study presents a novel modeling approach to study the sustainable production of CO_2-free hydrogen through non-plasma methane pyrolysis using microwave energy and carbon particles in a fluidized bed reactor (FBR). We have validated this technology through experimental means, with results showing that over 95% of methane is converted into hydrogen with a selectivity of over 90% and pure solid carbon as a by-product. In this technology, carbon particles, with an average size of 350 microns, in FBR absorb microwave energy and create a hot medium (>1200°C) in contact with flowing methane. As a result, methane decomposes into its constituents of matter, namely, hydrogen and solid carbon. Currently, the chemical and energy industry lacks a proper model that can simultaneously describe the nature of microwave heating and fluidization. Modeling such a complex phenomenon requires the multiphysics combination of electromagnetic (EM) simulations, fluid flow, and particle dynamics. To this end, the interaction of microwaves with carbon particles in the FBR is modeled by solving EM equations, otherwise known as Maxwell's equations, in a 1 kW-unit microwave cavity. Firstly, the FBR is assumed to be a uniform medium with a porosity of 0.4. The uniform medium approximation method divulges that the FBR of carbon particles only absorbs ~ 18 W out of 500 W of microwave input power, contradicting the absorbed power of ~ 100 W obtained from the experimental data. Hence, this finding emphasizes the fact that the FBR of carbon particles cannot be considered as a uniform medium due to their considerable electrical conductivity. Therefore, the microwave heating of the FBR is studied at a particle scale for the very first time. The simulation results reveal that electric charges are formed on each individual particle inside the FBR, contributing to the absorption of microwave energy mainly due to Joule heating rather than dielectric heating. A one-way coupling algorithm is then developed where the distribution of the electric field inside the FBR is imported into a CFD-DEM model. The developed CFD-DEM model will shed insights into the effect of fluidization mixing on the temperature distribution due to non-uniform microwave heating. Given this, the proposed model can serve as a valuable tool to study any chemical production processes that utilize electrification and fluidization, improving the efficiency of energy usage and reducing costs within the chemical production sectors.

AUTHOR INDEX